THINK TANK

EDITED BY DAVID J. LINDEN

THINK TANK

FORTY NEUROSCIENTISTS EXPLORE THE
BIOLOGICAL ROOTS OF HUMAN EXPERIENCE

Yale UNIVERSITY PRESS   NEW HAVEN AND LONDON

Published with assistance from the foundation established in memory of Calvin Chapin of the Class of 1788, Yale College.

Yale University Press books may be purchased in quantity for educational, business, or promotional use. For information, please e-mail sales.press@yale.edu (U.S. office) or sales@yaleup.co.uk (U.K. office).

Set in Scala & Scala Sans type by Integrated Publishing Solutions, Grand Rapids, Michigan. Printed in the United States of America.

Library of Congress Control Number: 2017955418
ISBN 978-0-300-22554-9 (hardcover : alk. paper)

A catalogue record for this book is available from the British Library.

This paper meets the requirements of ANSI/NISO Z39.48-1992 (Permanence of Paper).

10 9 8 7 6 5 4 3 2 1

Ornament: Computer artwork of a nerve cell, also called a neuron. Alfred Pasieka / Science Photo Library (adapted).

CONTENTS

RELATING

DECIDING

Scientists are trained to be meticulous when they speak about their work. That's why I like getting my neuroscience colleagues tipsy. For years, after plying them with spirits or cannabis, I've been asking brain researchers the same simple question: "What idea about brain function would you most like to explain to the world?" I've been delighted with their responses. They don't delve into the minutiae of their latest experiments or lapse into nerd speak. They sit up a little straighter, open their eyes a little wider, and give clear, insightful, and often unpredictable or counterintuitive answers.

This book is the result of those conversations. I've invited a group of the world's leading neuroscientists, my dream team of unusually thoughtful, erudite, and clear-thinking researchers, to answer that key question in the form of a short essay. Although I have taken care to invite contributors with varied expertise, it has not been my intention to create an informal comprehensive textbook of neuroscience in miniature. Rather, I have chosen a diverse set of scientists but have encouraged each author to choose her or his own topic to tell the scientific story that she or he is burning to share.

But let's face it: most books about the brain are not written by brain researchers, and most of them are not very good. Many are dull, and those that are readable are often uninformed or even fraudulent. This is the age of the brain, but thoughtful people have become understandably skeptical, having been inundated by a fire hose of neurobullshit ("looking at the color blue makes you more creative" or "the brains of Republicans and Democrats are structurally different"). I believe that readers hunger for reliable and compelling information about the biological basis of human experience. They want to learn what is known, what we suspect but cannot yet prove, and what remains a complete mystery about neural function. And they want to believe what they read.

The purpose of this book is not to launch a screed against neurobullshit but rather to offer an honest, positive recounting of what we know about the biology that underlies your everyday experience, along with some speculation about what the future will hold in terms of understanding the nervous system, treating its diseases, and interfacing with electronic devices. Along the way, we'll explore the genetic basis of personality; the brain substrates of aesthetic responses; and the origin of strong subconscious drives for love, sex, food, and psychoactive drugs. We'll examine the origins of human individuality, empathy, and memory. In short, we'll do our best to explain the biological basis of our human mental and social life and the means by which it interacts with and is molded by individual experience, culture, and the long reach of evolution. And we'll be honest about what is known and what is not. Welcome to the think tank!

David J. Linden
Baltimore, USA

THINK TANK

Primer

OUR HUMAN BRAIN WAS NOT DESIGNED
ALL AT ONCE BY A GENIUS INVENTOR
ON A BLANK SHEET OF PAPER

David J. Linden

THIS IS MY ATTEMPT to boil down the basic facts of cellular neuroscience into a small cup of tasty soup. If you've already studied neuroscience or you like to read about brain function, then you've likely heard much of this material before. I won't be offended if you skip this part of the meal. But if you haven't or if you're looking for a refresher, this section will serve to bring you up to speed and prepare you well for the essays that follow.

Around 550 million years ago it was simple to be an animal. You might be a marine sponge, attached to rock, beating your tiny whip-like flagella to pass seawater through your body in order to obtain oxygen and filter out bacteria and other small food particles. You'd have specialized cells that allow parts of your body to slowly contract to regulate this flow of water, but you couldn't move across the sea floor properly. Or you might be an odd, simple animal called a placozoan, a beast that looks like the world's smallest crepe—a flattened disc of tissue about 2 millimeters in diameter with cilia sprouting from your underside like an upside-down shag carpet. These cilia propel you slowly across the sea floor, allowing you to seek out the clumps of bacteria growing on the sea floor that are your food. When you found a particularly delicious clump, you could fold your body around it and secrete digestive juices into this makeshift pouch to speed your absorption of nutrients. Once digestion was finished, you would then unfold yourself and resume your slow ciliated crawl. Remarkably, as either a sponge or a placozoan, you could accomplish all sorts of useful tasks—sensing and responding to your environment, finding food, moving slowly, and reproducing yourself—without a brain or even any of the specialized cells called neurons that are the main building blocks of brains and nerves.

Neurons are wonderful. They have unique properties that allow them to rapidly receive, process, and send electrical signals to other neurons, muscles, or glands. The best estimates are that neurons first appeared about 540 million years ago in animals that were similar to modern-day jellyfish. We aren't sure why neurons evolved, but we do know that they appeared at roughly the same time that animals first started to eat each other, with all of the chasing and escaping that entails. So it's a reasonable hypothesis that neurons evolved to allow for more rapid sensing and movement, behaviors that became useful once life turned into a critter-eat-critter situation.

Neurons come in a variety of sizes and shapes, but they have many structures in common. Like in all animal cells, a thin, outer membrane encloses a neuron. Neurons have a cell body, which contains the cell nucleus, a storehouse of genetic instructions encoded in DNA. The cell body can be triangular, round, or ovoid and ranges in size from 4 to 30 microns across. Perhaps a more useful way to think about this size is that 3 typical neuronal cell bodies laid side by side would just about span the width of a human hair. Growing from the cell body are large, tapering branches called dendrites. These are the location where a neuron receives most of the chemical signals from other neurons. Dendrites can be short or long, spindly or shaggy, or, in some cases, entirely missing. Some are smooth while others are covered with tiny nubbins called dendritic spines. Most neurons have at least several branching dendrites, and they also have a single long, thin protrusion growing from the cell body. Called the axon, this is the information-sending part of the neuron.

Direction of Information Flow

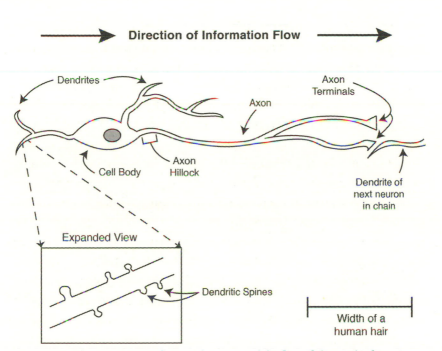

FIGURE 1. The major parts of a typical neuron and the flow of electrical information from one neuron to another.

While a single axon grows from the cell body, it often branches, and these branches can travel to various destinations. Axons can be very long. For example, some run all the way from a person's toes to the top of the spinal column.

Information is sent from the axon of one neuron to the dendrite of the next at specialized connections called synapses. At synapses, the tips of axons of one neuron come very close to, but do not actually touch, the next neuron (figure 1). The axon terminals contain many tiny balls made of membrane. Each of these balls, called synaptic vesicles, is loaded with about 1,000 molecules of a special type of chemical called a neurotransmitter. There is a very narrow saltwater-filled gap between the axon terminal of one neuron and the dendrite of the next called the synaptic cleft. On average, each neuron receives about five thousand synapses, mostly on the dendrites, with some on the cell body and a few on the axon. When we multiply 5,000 synapses per neuron by 100 billion neurons per human brain, the result is an enormous number as an estimate of the number of synapses in the brain: 500 trillion. To put this number in perspective, if you wanted to give away your synapses, each person on the planet (in 2017) could receive about 64,000 of them.

Synapses are the switching points between two forms of rapid signaling in the brain: electrical impulses and the release and subsequent action of neurotransmitters. The basic unit of electrical signaling in the brain is a rapid blip called a spike. Spikes are brief, large electrical events, about a millisecond or two in duration. They originate where the cell body and the axon join, at a spot called the axon hillock. The brain is bathed in a special saltwater solution called cerebrospinal fluid, which contains a high concentration of sodium and a much lower concentration of potassium. These sodium and potassium atoms are in their charged state, called ions, in which they each have one unit of positive charge. There is a gradient of sodium ion concentration across the outer membranes of neurons: the concentration of sodium ions outside a neuron is about fifteenfold higher than it is inside. The gradient for potassium runs in the other direction: the concentration of potassium ions is about fiftyfold higher inside than outside. This situation is crucial for the electrical function of the brain. It creates potential energy, similar to winding the spring on a child's toy, and the energy can then be released in the appropriate circumstances to create electrical signals in neurons. Neurons rest with an electrical potential across their outer membranes: there is more negative charge inside than out. When a spike is triggered, specialized doughnut-shaped proteins embedded in the outer membrane, called sodium channels, open their previously closed doughnut hole to let sodium ions rush in. A millisecond or so later, a different kind of ion channel, one that passes potassium ions, opens up, allowing potassium to rush out, thereby rapidly terminating the spike.

Spikes travel down the axon to the axon terminals, and when they arrive there, they trigger a series of chemical reactions. These chemical reactions cause synaptic vesicles to fuse with the outer membrane of the axon terminal, releasing their contents, including neurotransmitter molecules, into the synaptic cleft. The released neurotransmitter molecules then diffuse across the narrow synaptic cleft to bind neurotransmitter receptors, which are embedded in the outer membrane of the next neuron in the signaling chain. One form of neurotransmitter receptor, called an ionotropic receptor, is like a closed doughnut that only opens its hole when it is bound by neurotransmitters. If the ion channel in that receptor allows positive ions to flow in, then this excites the receiving neuron. Conversely, if the ion channel opened by the neurotransmitter allows positive ions to flow out of the neuron (or negative ions like chloride to flow in), this will inhibit spike firing in the receiving neuron.

Electrical signals from activated receptors at synapses all over the dendrite and cell body flow toward the axon hillock. If enough excitatory electrical signals from the synapses arrive together and they are not negated by simultaneous inhibitory signals, then a new spike will be triggered there, and the signal will be passed down the axon of the receiving neuron. Most of the psychoactive drugs that we consume, both therapeutic and recreational, act at synapses. For example, sedatives like Xanax and related compounds work by enhancing inhibitory synapses and in this way reducing the overall rate of spike firing in certain regions of the brain.

Electrical signaling in the brain is fast by biological standards (in the range of milliseconds), but this signaling is still about a millionfold slower than the electrical signals coursing through the circuits of your laptop computer or smartphone. It is important that not all signaling at synapses is fast. In addition to the ionotropic neurotransmitter receptors that work on the timescales of milliseconds, there is a much slower group called metabotropic receptors. These receptors do not have an ion channel pore as part of their structure, but rather trigger or block chemical reactions in the receiving neuron and act on a timescale of seconds to minutes. The fast ionotropic receptors are useful for rapid signals like those that convey visual information from your retina to your brain or carry commands from your brain to your muscles to undertake a voluntary movement. By contrast, the slow metabotropic receptors, which respond to neurotransmitters including serotonin and dopamine, are more often involved in determining your overall state of mind like your alertness, mood, or level of sexual arousal.

A single neuron is almost useless, but groups of interconnected neurons can perform important tasks. Jellyfish have simple nets of interconnected neurons that allow them to adjust their swimming motions to respond to touch, body tilt, food odors, and other sensations. In worms and snails, the cell bodies of neurons have become clustered into groups called ganglia, and these ganglia are interconnected by nerves that are cables consisting of many axons bound together. Ganglia in the head have fused together to form simple brains in lobsters, insects, and octopuses. The octopus brain contains about 500 million neurons, which seems like a large number but is only about 1/200th of the size of the human brain. Nonetheless, an octopus can perform some impressive cognitive feats. For example, it can watch another octopus slowly solve a puzzle box to get food hidden inside and then apply that learning to immediately open the puzzle box when given access to it for the first time. As vertebrate evolution has proceeded, from frogs to mice to monkeys to humans, brains have mostly gotten bigger (relative to body size), and the neurons within have become more interconnected with each other, with the largest expansion occurring in the neocortex, the outermost portion of the brain.

The evolution of brains or any other biological structures is a tinkering process. Evolution proceeds in fits and starts with lots of dead ends and errors. Most important, there's never a chance to wipe the slate clean and do a totally new design. Our human brains were not designed all at once, by a genius inventor on a blank sheet of paper. Rather, the brain is a pastiche, a grab bag of make-do solutions that have accumulated and morphed since the first neurons emerged. It is a cobbled-together mess that nonetheless can perform some very impressive feats.

That the design of the human brain is imperfect is not a trivial observation; suboptimal brain design deeply influences the most basic human experiences. The overall design of the neuron hasn't changed very much since it first emerged, and it has some serious limitations. It's slow, unreliable, and leaky. So to build clever humans from such crummy parts, we need a huge interconnected brain with 500 trillion synapses. This takes a lot of space—about 1,200 cubic centimeters (cc). That's so big that it would not fit through the birth canal. Changes to the pelvis to make a larger birth canal would presumably interfere with upright walking. So the painful solution is to have human babies born with 400-cc brains (about the size of an adult chimpanzee's brain). Even this size is still a problem—the baby's head barely fits through the vagina. (In fact, death in childbirth, while common through most of human history, is almost unheard of in other mammals.) Once born, humans undergo an unusually long childhood while that 400-cc brain matures and grows, a process that is not complete until about age twenty. There's no other animal species in which an eight-year-old cannot live without its parents. Our extra-long human childhoods drive many aspects of human social life, including our dominant mating system of long-term pair bonding, an aspect that is very rare in the mammalian world. Or to put it another way, if neurons could have been optimally redesigned at some point in evolution, we likely wouldn't have marriage as a dominant cross-cultural institution.

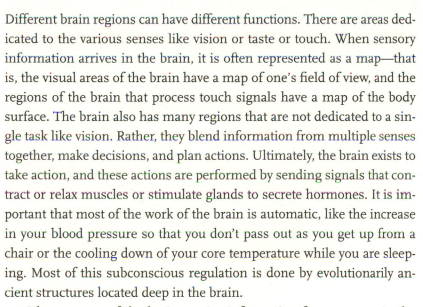

Different brain regions can have different functions. There are areas dedicated to the various senses like vision or taste or touch. When sensory information arrives in the brain, it is often represented as a map—that is, the visual areas of the brain have a map of one's field of view, and the regions of the brain that process touch signals have a map of the body surface. The brain also has many regions that are not dedicated to a single task like vision. Rather, they blend information from multiple senses together, make decisions, and plan actions. Ultimately, the brain exists to take action, and these actions are performed by sending signals that contract or relax muscles or stimulate glands to secrete hormones. It is important that most of the work of the brain is automatic, like the increase in your blood pressure so that you don't pass out as you get up from a chair or the cooling down of your core temperature while you are sleeping. Most of this subconscious regulation is done by evolutionarily ancient structures located deep in the brain.

The neurons of the brain receive information from sensors in the eyes, ears, skin, nose, and tongue (and other places too). Moreover, sensory information doesn't come just from detectors that point outward at the external world but also from those that point inward to monitor such aspects as the tilt of your head or your blood pressure or how full your stomach is. Within the brain, neurons are highly interconnected with each other. Crucially, all of this wiring, consisting of axons that run from place to place, must be specific: signals from the retina need to go to the vision-processing parts of the brain, commands from the motion-producing parts of the brain must ultimately make their way to muscles, etc. If mistakes are made and the brain is mis-wired, even subtly, then all sorts of neurological and psychiatric problems can result.

How does this specific brain-wiring diagram become established?

The answer is that it is determined by a mixture of genetic and environmental factors. There are genetic instructions that specify overall shape and the wiring diagram of the nervous system on the large scale. But in most locations the fine-scale neural wiring must be refined by local interactions and experience. For example, if a baby is born but its eyes remain closed in early life, then the visual parts of its brain will not develop properly and it will not be able to see, even if the eyes are opened in adulthood. When the brain is developing, in utero and through early life, about twice as many neurons are created than are ultimately used, and many synapses are formed and later destroyed. Furthermore, those synapses that are formed and retained can be made weaker or stronger as a result of experience. This process, by which experience helps to form the brain, is called neural plasticity. It is important in development, but it is also retained in an altered form in adulthood. Throughout life, experience, including social experience, fine-tunes the structure and function of the nervous system, thereby creating memories and helping to form us as individuals.

Science Is an Ongoing Process, Not a Belief System

William B. Kristan, Jr., and Kathleen A. French

ONE OF THE MOST DIFFICULT IDEAS to explain to the general public is what it means to "believe in" a scientific concept. In part, this difficulty arises because the word "believe" can have different meanings. In our daily lives, we use "believe" in many contexts:

I believe it will rain soon.
I believe my child when (s)he says that (s)he doesn't use recreational drugs.
I believe that the defendant is guilty.
I believe that the cerebral cortex is the site of consciousness.
I believe that A will make a better president than B.
I believe in gravity.
I believe in God.

In some of these examples, "I believe" means "I am certain of," whereas in other examples, it means something like "I hold an opinion" or "I suppose," as in the speculation about the possibility of rain. In all cases, the believer may well take action based upon the belief, and the action might be as trivial as grabbing an umbrella before heading outdoors or as far-reaching as basing one's life on religious teachings. Where does belief in a scientific concept fit into this spectrum? This question is difficult to answer because there are different stages in the development of scientific concepts, with widely different criteria for judging them. These stages arise because science uses a guess-test-interpret strategy, and this sequence is typically repeated many times. In fact, in everyday life, we all act like scientists—at least sometimes.

Consider a real-life example. You sit down in your favorite chair to read the newspaper and flip on the switch for your reading lamp, but the lamp fails to light. Maybe someone unplugged the cord *(guess 1)*. You look at the wall, but the cord remains plugged into its socket *(test 1)*, so that's not the problem *(interpretation 1)*. Maybe the circuit breaker was opened: a reasonable *guess 2*, but the TV—which is on the same circuit—is working *(test 2)*, so it's not a circuit-breaker problem *(interpretation 2)*. Perhaps the problem is in the wall socket *(guess 3)*, so you plug another lamp into it, and that one works just fine *(test 3)*, so the wall socket is functioning properly *(interpretation 3)*. You work your way through successive guesses (bulb? broken cord?) and tests to arrive at an interpretation (bad lamp switch) that ultimately enables you to fix the lamp. Previous experiences with circuit breakers, wall sockets, and lamps, and a rough understanding of electrical currents, informed your guesses.

In its basic logic, doing science isn't so different from fixing your lamp, except that each step may be more complex. One approach—which started with Aristotle—is inductive: you gather all the facts you can about a specific topic, think hard, and then insightfully conclude ("induce") the general relationship that explains the facts.[1] This approach is common, and it has produced explanations both sacred (e.g., creation stories) and mundane (e.g., trying to decide why your car won't start). As experimental science blossomed in the past century or two, however, the value of this inductive technique has transformed from being the source of an ultimate explanation to formulating a guess. (Scientists like the term "hypothesis," philosophers seem to prefer "conjecture," but both are essentially synonyms for "guess.")[2]

So has guessing become a trivial and unimportant part of doing science? Far from it! Good guesses require both a lot of background knowledge and great creativity. Typically, a good guess is at least somewhat surprising (no one else has either thought of it or has dismissed it), is broadly interesting, is testable, and holds up under many tests. Sometimes the term "falsifiable" is used instead of "testable"—that is, for a guess to qualify as scientific, it must be vulnerable to falsification by objective, repeatable tests.[3] The kinds of tests required to evaluate a hypothesis (i.e., to accept or reject the guess) are stringent. (Accepting a hypothesis means that it has not yet been rejected.) Reduced to its simplest level, science attempts to find causal relationships, so a scientific guess typically has the form "A *causes* B." Here is an example from our laboratory's study of the medicinal leech. We guessed that some neurons in the leech nervous system activated its swimming behavior. Based on her initial experiments, Janis Weeks, a graduate student, found a type of neuron that seemed to fit that role; she named it cell type 204.[4] How could we test her guess that cell type 204 caused swimming? In general, there are three common categories of tests for causality: correlation, necessity, and sufficiency. Janis's experiments with cell 204 employed all three categories.

Correlation. Electrical recordings from cells 204 showed that they were always active just before and continued throughout the time that the animal swam—that is, the cells' activity was *correlated* with swimming. Note that even this weakest test of causality could have falsified our guess if cell 204 was not active during swimming. In other words, tests of correlation can disprove a guess but cannot prove it.

Sufficiency. Stimulating a single cell 204 (one of the approximately 10,000 neurons in the leech's central nervous system) caused the animal to swim. We concluded that activating a single cell 204 is *sufficient* to cause a leech to swim. But this test could not show that activating cell 204 was the only way to induce swimming. Janis needed to do further tests.

Necessity. Inactivating a single cell 204 (by injecting inhibitory electric current into it) reduced the likelihood that stimulating a nerve would cause swimming, showing that activity in cell 204 was at least partially *necessary* for swimming. (There are twelve cells 204 in the leech nervous system and only two of them could be controlled at a time, a factor that explains the reduction in—but not total blocking of—swimming.)

Based on these results, and similar ones from other nervous systems, neurons like cell 204 have been called "command neurons" because their activity elicits ("commands") a specific behavior. The notion is that command neurons link sensory input with motor parts of the brain: they get input from sensory neurons, and if this input activates them, they initiate a specific motor act. Such neurons have also been called "decision makers," an implicit guess that their true function is to make a choice between one behavior (swimming) and other behaviors (e.g., crawling).

The basic experiments on cells 204 were performed nearly forty years ago, so we can ask the following: do we still believe the original guess-test-interpretation story?[5] The answer is yes and no. The basic data have stood the test of time (and many repetitions), but further experiments have uncovered additional neurons that produce results similar to those of cells 204, so our initial conclusion that cells 204 were uniquely responsible for swimming was too simple. In further experiments using dyes that glow to report electrical activity, which allowed us to monitor the activity of many neurons at once, it became clear that subtle interactions among many other neurons acting together decide whether a leech swims or crawls. Cells 204, along with the additional "command neurons," carried out the motor behavior once these subtle interactions ended. So cell 204 is not a "commander-in-chief" but something more like a lieutenant who puts into action the commands issued by the joint chiefs, who actually make the decision.[6]

Remembering the experiments on cell 204, we return to the meaning of "belief" in science. Minimally, this question needs to be broken into at least three different levels:

1. Can the guess be falsified? If there is no way to falsify a guess by using objective, real-world tests, it can be interesting, but it falls outside the realm of science.

2. Do we trust the validity of the data? To answer this question, we must consider whether the techniques used were appropriate, whether the experiments were done with care, and whether the results are convincing. For instance, in a typical experiment intended to elucidate the function of a region of the brain, the function of that area will be experimentally modified, and experimenters will look for a change in behavior and/or brain activity. In looking for change, the experimenter applies a stimulus and scores the response. Often the data are messy: maybe when the same stimulus is repeated, it elicits a variety of responses, or two different stimuli may elicit the same response. A number of issues can cause such a result, and there are established ways to identify

and solve these problems. For example, the person who evaluates the results is prevented from knowing the details of the treatment (it is called "blinding" the experimenter). Alternatively, the experiment may be repeated in a different laboratory, so the equipment, people, and culture of the laboratory are different.

3.   Do we believe the interpretations? In general, an interpretation is the most interesting part of any scientific study (and it is the part most likely to be carried in the popular press), but it is also the most subject to change. As shown by the findings about cell 204 in the leech nervous system, new data can change the interpretation considerably, and that process is continuous. Karl Popper, an influential twentieth-century philosopher of science, argued that science cannot ever hope to arrive at ultimate truth.[7] A well-founded current estimate of truth can explain all—or at least most—of the current observations, but additional observations will eventually call into question every interpretation, replacing it with a more comprehensive one. He argues that this process does not negate the old interpretation, but rather the new data provide a closer approximation to ultimate truth. In fact, the interpretations of one set of data generate the guesses for the next set of experiments, just as you found in repairing your faulty reading lamp.

So how does "scientific belief" differ from other sorts of belief? One major difference is that science—at least experimental science—is limited only to ideas that can be tested objectively, reproducibly, and definitively; if others do exactly the same experiments, they will get the same results. This qualification eliminates from scientific inquiry a large number of deeply interesting questions, such as "Why am I here?" and "Is there a Supreme Being?" These qualifications even eliminate whole disciplines, such as astrology, that act like science in that they gather huge amounts of data but whose conclusions cannot be objectively tested.[8] Scientific papers usually separate "results" from "the discussion." Belief in the results requires judging whether the experiments were done properly and whether other scientists can reproduce the findings; such judgments are relatively objective. Believing what is said in the discussion section is more nuanced: Do the data support the interpretation? Are the conclusions reasonable, based upon the results in this and previous papers? Does the interpretation point to further testable guesses? The discussion, although often the most interesting part of any scientific paper, is also the part that is least likely to stand the test of time. To someone outside the field of study, the changes in interpretations can be confusing and frustrating (e.g., Is fat in my diet good or bad for me?), but these successive approximations are inherent in the process. The interpretations are where the poetry lies, where creativity is most obvious in science. The fact that interpretations change, however, means that all statements of belief carry an inherent asterisk: what a scientist believes today can change greatly with the next set of experiments that he or she does or—less happily— that another scientist does. Scientists must be able to let go of their fondest beliefs and adopt new points of view when data require it, and nonscientists need to understand the dynamic nature of these beliefs.

NOTES

1. Inductive reasoning as the best model for scientific thought had a remarkably long run, involving many great philosophers, including—in addition to Aristotle—David Hume (*Treatise of Human Nature;* London: Thomas and Joseph Allman, 1817), Immanuel Kant (*Critique of Pure Reason;* New York: Colonial Press, 1899), and John Stuart Mill (*A System of Logic;* London: John W. Parker, 1843). They argued that the job of a scientist was first to collect data about a topic of interest without thinking about the relationship among the pieces of information collected (because thinking about cause and effect might bias the data gathering), and then, in a blinding flash of insight, the answer would become clear. Twentieth-century philosophers like Karl Popper (*Conjectures and Refutations;* New York: Routledge, 1963) and Thomas Kuhn (*The Structure of Scientific Revolutions;* Chicago: University of Chicago Press, 1962) argued that true induction is barren (it is limited to the data at hand), and, anyway, scientists don't operate in that manner. Instead, they have at least a vague idea (a guess) about what is important, and that idea guides which data are crucial to be gathered, whereupon the iterative guess-test-interpret cycle kicks in. A wonderful discussion of this topic is in P. B. Medawar's essay, "The Philosophy of Karl Popper" (1977), in his posthumously published book of essays entitled *The Threat and the Glory* (New York: Harper Collins, 1990). Every nascent scientist should be required to read this essay for inspiration before stepping into a laboratory, and every nonscientist should read it twice for clarity.

2. These formulations have also been called "happy guesses" or, more pompously, "felicitous strokes of inventive talent"; we'll stick with "guesses." The "felicitous strokes" quote is from William Whewell, *History of the Inductive Sciences* (London: John W. Parker, 1837), cited in an enlightening book by P. B.

Medawar, *The Limits of Science* (New York: Harper and Row, 1984). Sir Peter Medawar was an extremely successful British immunologist (he was awarded the Nobel Prize in Physiology or Medicine in 1960) who wrote many essays and books for the general public on science and also on philosophy for scientists. They are models of clarity and a delight to read.

3. There are many books on this topic. A definitive treatment is Karl Popper's *The Logic of Scientific Discovery*, first published in German in 1935 and translated into English in 1959. It is readily available through Routledge Classics (New York, 2002), although it is a bit dense. (It is considered a bit old-fashioned by philosophers, but research scientists find it captures much of what they do every day.) A more accessible, more modern book on a similar topic is *Failure*, by Stuart Firestein (New York: Oxford University Press, 2016).

4. J. C. Weeks and W. B. Kristan, Jr., "Initiation, Maintenance, and Modulation of Swimming in the Medicinal Leech by the Activity of a Single Neuron," *Journal of Experimental Biology* 77 (1978): 71–88.

5. The latest review of the circuitry underlying leech swimming is in the following review article: W. B. Kristan, Jr., R. L. Calabrese, and W. O. Friesen, "Neuronal Basis of Leech Behaviors," *Progress in Neurobiology* 76 (2005): 279–327.

6. In recent years, it has become possible to do similar experiments in brains more like our own than leech brains are, allowing scientists to ask whether there are neurons in our own brains that act the way cells 204 act in the leech. The huge number of neurons in the brains of mammals has been a challenging obstacle for addressing questions about the functions of individual neurons. However, over the past decade, techniques for imaging and for selectively expressing activity-reporter molecules in neurons have allowed the study of the behavioral functions of many neurons at once. As a result, the same sorts of experiments described for cell 204 are now possible in more complex brains, such as those of fish and mice. While scientists image neurons of known function, their activity can be correlated with behavior. By genetically manipulating the neurons to produce light-sensitive proteins, then stimulating them with appropriately colored lights, these neurons can be turned on or off during the performance of a behavior to test for their sufficiency and necessity in causing that behavior. Such experiments, among others, are revolutionizing the study of the neuronal basis of behaviors in animals with complex brains. For detailed information about this approach, the following references are a good place to begin. The first two emphasize the technique itself, and the last two address the kinds of experiments that are being done with the technique: K. Deisseroth, "Controlling the Brain with Light," *Scientific American* 303 (2010): 48–55; K. Deisseroth, "Optogenetics: 10 Years of Microbial Opsins in Neuroscience," *Nature Neuroscience* 18 (2015): 1213–1225; E. Pastrana, "Primer on Optogenetics. Optogenetics: Controlling Cell Function with Light," *Nature Methods* 8 (2010): 24–25; V. Emiliani, A. E. Cohen, K. Deisseroth, and M. Haeusser, "All-Optical Interrogation of Neural Circuits," *Journal of Neuroscience* 35 (2015): 13917–13926.

7. Karl Popper wrote about the notion of successive approximations to truth in several places, but the most accessible is in an essay entitled "Science: Conjectures and Refutations," which was originally given as a lecture in Cambridge, England, in 1953 and published in his book *Conjectures and Refutations: The Growth of Scientific Knowledge* (London and New York: Routledge, 1963). This essay is available online at http://worthylab.tamu.edu/courses_files/popper_conjecturesandrefutations.pdf.

   Stuart Firestein, in *Failure*, agrees that a rejection of guesses is the usual way that scientific progress is made, but he makes the argument from a somewhat different perspective. Firestein argues that few scientists actually follow the guess-test-interpret strategy on a day-to-day basis, although they write their research papers as though they do. He calls the rejection of a guess one type of failure and makes the case that this type of failure is the most beneficial for the progress of science.

8. Karl Popper revealed in *The Logic of Scientific Discovery* that he became interested in philosophy by wondering how science is different from such diverse areas of knowledge as astrology, metaphysics, and psychoanalysis. He concluded that the distinction between science and nonscience lies in the testability and refutability of scientific theories: "Every 'good' scientific theory is a prohibition; it forbids certain things to happen. The more a theory forbids, the better it is" ("Science: Conjectures and Refutations"). (Firestein, in *Failure*, adds to this list some more modern topics like Scientology, intelligent design, and many alternative medicine treatments as other examples of theories that do not lend themselves to refutability.) Popper pointed out that there are perfectly good areas of intellectual pursuit—like metaphysics and ethics—that are critically important for human culture and survival but that are inherently nonscientific because they are not testable. This issue was nicely addressed by Sir Peter Medawar in *The Limits of Science* in two pithy quotes:

   If the art of politics is indeed the art of the possible, then the art of scientific research is surely the art of the soluble. (P. 21)

   It is not to science, therefore, but to metaphysics, imaginative literature or religion that we must turn for answers to questions having to do with first and last things [e.g., "What is the point of living?"]. Because these answers neither arise out of nor require validation by empirical evidence, it is not useful or even meaningful to ask whether they are true or false. The question is whether or not they bring peace of mind in the anxiety of incomprehension and dispel the fear of the unknown. (P. 60)

DEVELOPING, CHANGING

Genetics Provides a Window on Human Individuality

Jeremy Nathans

ANYONE WHO HAS SPENT TIME in a room full of four-year olds has seen the evidence. Even at a young age, we humans show striking personality differences. Some children are outgoing; others are shy. Some children are focused; others jump from one activity to another. Some children are strong-willed; others, less so. Personality traits largely define who we are as adults—pessimistic or optimistic, sociable or solitary, authoritarian or free-spirited, empathetic or suspicious. Aggregated over thousands or millions of people, such traits define the characteristics of our societies.

What determines personality? To what extent is it innate? To what extent is it molded by experience? At their core, these are questions about brain development, function, and plasticity, and they are some of the deepest questions that we can ask of brain science.

More than one hundred years ago, the British polymath Francis Galton posed these questions in essentially their modern form.[1] Galton conceptualized the forces that mold personality, intelligence, and other mental characteristics as a reflection of the combined contributions of "nature and nurture." Over the past century, research in animal behavior, psychology, and genetics has begun to converge and to focus this inquiry.

In a consideration of the lessons that we might glean from our nonhuman relatives, there is no better place to start than with the work of

Galton's cousin, Charles Darwin, who was fascinated by the changes in physical appearance and behavior that could be elicited by the selective breeding of domesticated animals. Consider, for example, the personalities of dogs. As every dog owner knows, individual dogs have distinctive temperaments, skills (or lack thereof), and habits—in short, a set of traits that defines the dog's personality. Strikingly, these attributes have a strong genetic component. A golden retriever's warm personality, an Australian sheepdog's herding instinct, and a German shepherd's self-discipline are, in large part, the product of selective breeding. To dog owners and breeders, these behavioral traits are as valued and as distinctive as the dogs' physical features.

If we examine the broadest of canine behavioral traits—those that distinguish wild and domesticated dogs—we observe that the critical characteristic shared by all domesticated breeds is tameness, a fundamental change in the ground rules of interpersonal interactions with humans. This trait is exemplified by a dramatic change in the meaning of eye contact, from threat to affection. In a landmark study of Siberian silver foxes conducted by Dmitri Balyaev, Lyudmila Trut, and their colleagues, the behavioral transition from a wild to a tame temperament was achieved with only 30–40 generations of selective breeding of wild foxes.[2] This breeding program, carried out in Novosibirsk starting in the late 1950s, ultimately produced foxes that exhibited many of the endearing traits that we associate with domesticated dogs, including tail wagging, hand licking, responding to human calls, and a desire for physical and eye contact with humans.

One of the lessons from the Novosibirsk study is that the underlying genetic variation required for the transition from a wild to a tame temperament preexisted in the wild fox population. Indeed, the researchers observed that "friendly" behavior began to emerge after only four generations of breeding ("friendliness" being defined by the interactions between foxes and humans). At present, the precise genetic changes responsible for tameness in Siberian foxes are not known, but Balyaev, Trut, and their colleagues have presented evidence that—whatever those changes are—they lead to hormonal changes that include a lowering of levels of stress hormones, such as glucocorticoids. Perhaps a "type A" personality is optimally suited to a world in which the next meal is unpredictable and every large animal is a likely adversary.

To what extent do these insights into the genetic control of behavioral traits in animals apply to us? In 1979, Thomas Bouchard, a psychologist at the University of Minnesota, launched one of the most ambitious attempts to answer this question. Over the next twenty years, Bouchard and his colleagues searched for those rare twins who had been adopted into different households and raised separately to determine the extent to which their psychological similarities and differences reflected shared genetics or different environments.[3] The Minnesota Study of Twins Reared Apart (MISTRA) compared identical twins and fraternal twins, and, in collaboration with Bouchard's University of Minnesota colleague David Lykken, it also compared twins reared apart with twins reared together.[4]

Identical twins (also called monozygous twins) arise from a single fertilized egg that, early in development, divides to form two embryos. The two individuals inherit the same version of each gene from their parents and are, therefore, genetically identical. Since identical twin embryos also inherit the same arrangement of X- and Y-chromosomes, they are also the same sex. That is, identical twin pairs can consist of two boys or two girls but never a boy and a girl. Approximately 1 out of every 270 humans is a member of an identical twin pair.

In contrast, fraternal twins (also called dizygous twins) arise when two eggs are released during the same ovulatory cycle, are independently fertilized by two sperm, and then develop into two embryos. The two individuals are only as similar to each other as are any other pair of siblings. The distinguishing feature of fraternal twin siblings, as compared to other siblings, is that these siblings share the same uterus and are the same age. Geneticists loosely say (or write) that "fraternal twins share, on average, 50 percent of their genes."[5] Similarly, since fraternal twin embryos have independently inherited their X- and Y-chromosomes, they are as likely to be the same sex (boy + boy or girl + girl) as opposite sex (boy + girl or girl + boy). Approximately 1 out of every 115 humans is a member of a fraternal twin pair.

The simplest twin study design is one in which the values for some quantifiable trait—for example, height, weight, or blood pressure—are determined for both members of a large number of fraternal and identical twin pairs. The differences in these values are calculated for each pair, and the results are compared between fraternal and identical groups. Since identical twins are always of the same sex, the study design limits the participating fraternal twins to those that are also of the same sex. One study of this type has shown, for example, that the average height difference between fraternal twins is approximately 4.5 centimeters, whereas the average difference between identical twins is approximately 1.7 centimeters. The smaller average difference between identical twins is attributed to their greater degree of genetic similarity.

The alert reader may have discerned a potential fly in the ointment for this type of study, especially as applied to traits with a psychological component. Identical twins often look so similar that they are confused with one another—an occasional source of amusement. As a result, they may find that their teachers, friends, or even relatives tend to treat them in similar ways, either because they cannot tell the twins apart or because they subconsciously assume that two people who look so much alike are also similar in other respects. A similarity in interpersonal interactions of this type creates what behavioral geneticists call a "shared environment," and it can confound the analysis of nature versus nurture. Additionally, as described below, identical twins do, in fact, tend to resemble one another on a wide range of personality traits and, perhaps in consequence, often develop an extraordinarily close bond with one another. This development leads to a second conundrum: maybe the close interpersonal relationship between many identical twins tends to reinforce their psychological similarities and suppress their differences.

Studying twins reared apart from birth or infancy neatly solves such problems. As MISTRA showed, a comparison between identical twins reared apart and fraternal twins reared apart is particularly informative. In this comparison, the twin pairs are either 100 percent genetically identical or on average 50 percent genetically identical, respectively, but their rearing environments are largely uncorrelated. Two other useful comparisons are between identical twins reared together versus apart and between fraternal twins reared together versus apart. The latter comparisons provide another approach to assessing the influence of shared versus unshared environment during the formative years of childhood.

Over its twenty-year life, the MISTRA scientists studied eighty-one pairs of identical twins raised apart and fifty-six pairs of fraternal twins raised apart. At the time that they were studied, the twins averaged forty-one years of age. They had spent an average of only five months together before being separated and then had no contact with one another for an average of thirty years. During the study, each participant typically spent a week at the University of Minnesota and underwent a comprehensive set of physical, medical, and psychological tests.

The results from the psychological tests were striking. Across a range of personality features, such as extraversion/introversion and emotional lability/stability, genetic influences were substantial, averaging roughly 40 percent of the variance, a statistical measure of the variation across the population being studied. Moreover, vocational interests and specific social behaviors, such as religiosity and traditionalism, showed a similarly large genetic influence.

The single most intensively studied of all human psychological traits is the Intelligence Quotient (IQ), which is determined by performance on a written test of knowledge and intellectual skill. Although the name "IQ" is unduly grandiose, the test is of both theoretical and practical interest as its results are strongly predictive of educational and vocational success.[6] The data from MISTRA showed that for the population studied, about 70 percent of the variance in IQ test scores could be explained by genetics. In particular, the difference in IQ test scores between identical twins raised apart was only slightly greater than the average difference between two test scores obtained when the same person took the test on two separate occasions. A systematic analysis of the different adopted households in which the identical twins were raised showed little influence on IQ test scores of parental educational level or household cultural or scientific enrichment. These results are remarkable, but they need a few qualifiers because they do not address the influence of extremes of environmental enrichment or deprivation. Nearly all of the MISTRA participants were raised in households and in communities that provided opportunities for a solid education, and therefore the strength of the genetic influence applies to that broadly permissive environment.

I have emphasized MISTRA because of its large size and exemplary design, but many dozens of other twin and family studies provide data on personality and IQ that closely agree with the results from MISTRA.[7] An especially intriguing comparison of 110 identical twin pairs and 130 fraternal twin pairs who were all over eighty years old found a higher degree of similarity for identical twins compared to fraternal twins on every measure tested, including general cognitive ability (IQ), memory, verbal ability, spatial ability, and processing speed.[8] This study also showed that the genetic influence on IQ does not decline appreciably with age, although this and other studies do not address the extent to which a person's IQ score reflects motivation, curiosity, and self-discipline, in addition to "intelligence" per se.

Twin studies can measure the average contribution of genetics to individual variation in personality and cognitive ability, but they cannot reveal the biological mechanisms responsible for this variation. We can think of twin studies as providing us with data that are analogous to the performance characteristics of different types of automobiles. We may learn that a Porsche accelerates more rapidly than a Toyota, but to understand the cause of that difference we will need to know in detail how these two automobiles differ. We will also need to know a lot about how automobiles work in general.

In biology, looking under the hood means understanding how cells grow, become specialized, interact, and carry out their particular functions. It also means understanding how the genetic blueprint that each of us inherits codes for the proteins that comprise the molecular machines that underpin all cellular structure and function. This is a tall order, and we are still far from having a fully satisfactory understanding of these processes. However, progress over the past fifty years has been impressive. The fundamental mechanisms underlying communication between nerve cells are now known, as are many of the mechanisms responsible for assembling connections between nerves cells during development.

Progress in genetics has been especially rapid. We now have the complete DNA sequences of our species and many dozens of other species, and partial DNA sequences have been determined from hundreds of thousands of individual humans. These sequences show that our genetic blueprint is remarkably similar to the genetic blueprints of other mammals. Thus the large physical and mental differences among different mammalian species likely arise from the sum of many relatively subtle differences in gene structure and function. Additionally, a comparison of the DNA blueprints from different humans shows that we differ genetically from one another, on average, by only one part in one thousand.[9] Determining how these genetic differences contribute to making each of us who we are is one of mankind's greatest scientific challenges.

NOTES

1. F. Galton, *Inquiries into Human Faculty and Its Development* (London: J. M. Dart, 1907), available at www.galton.org.
2. D. K. Balyaev, "Destabilizing Selection as a Factor in Domestication," *Journal of Heredity* 70 (1979): 301–308; L. Trut, "Early Canid Domestication: The Farm-Fox Experiment," *American Scientist* 87 (1999): 160–169.
3. T. J. Bouchard, D. T. Lykken, M. McGue, N. L. Segal, and A. Tellegen, "Sources of Human Psychological Differences: The Minnesota Study of Twins Reared Apart," *Science* 250 (1990): 223–228.
4. N. L. Segal, *Born Together—Reared Apart: The Landmark Minnesota Twin Study* (Cambridge, MA: Harvard University Press, 2012).
5. The statement "fraternal twins share, on average, 50 percent of their genes" is really a compressed version of the more precise statement that "fraternal twins have a 50 percent chance of inheriting the same version of any particular gene from each parent." For example, if a mother has two different versions of a gene for eye color, then each of her children has a 50 percent chance of inheriting version one and a 50 percent chance of inheriting version two. From this statement, the reader can easily write down all of the possible combinations of inherited versions and see that, on average, fraternal twins inherit the same version half the time and different versions the other half of the time.
6. It is important to emphasize that, like all standardized testing, the results of IQ testing should be interpreted in a cultural context. Assuming that we could agree that "innate" intelligence exists and can be measured, it is unlikely that any standardized test could measure it in a manner that is free of cultural bias.
7. L. J. Eaves, H. J. Eysenck, and N. G. Martin, *Genes, Culture, and Identity: An Empirical Approach* (New York: Academic Press, 1989).
8. G. E. McLearn, B. Johnasson, S. Berg, N. L. Pederson, F. Ahern, S. A. Petrill, and R. Plomin, "Substantial Genetic Influence on Cognitive Abilities in Twins 80 or More Years Old," *Science* 276 (1997): 1560–1563.
9. E. S. Lander, "Initial Impact of the Sequencing of the Human Genome," *Nature* 470 (2011): 187–197.

Though the Brain Has Billions of Neurons, Wiring It All Up May Depend upon Very Simple Rules

Alex L. Kolodkin

THE IMMENSE COMPLEXITY of neural connections begs the following question: What labels, or cues, could possibly provide a code that instructs their precise organization? Imagine the task of connecting the thousands of phones in the new One World Trade Center building in New York City to their central switching stations—color-coded wires, numbered phone jacks, and lots of unique labels are the only hope to get it right. But to use this "unique label" strategy to wire up the human brain one would need trillions of specific molecular cues. Is such a wiring code even possible? Over one hundred years of neuroscience research has yielded only a few hundred molecules that are known to selectively direct the formation of connections among neurons. But even if all the genes in the human genome produced only wiring cues, that would result in approximately 20,000 unique cues, far fewer than necessary to uniquely code for all the connections in the human brain.[1] Recent work in the insect visual system shows that extremely complex neuronal connections among a very large number of neurons can be instructed by very simple rules; each individual neuron can follow these rules on its own and, in the absence of myriad unique labels, wire up intricate and specific connections to many other neurons. So to what extent can a nervous system self-assemble? The answer is, surprisingly, quite a bit.

Some of the greatest contributions to our understanding of both the complexity and logic of neural connections were made early in the last century by the Spanish neuroanatomist Santiago Ramón y Cajal.[2] Using microscopes primitive by today's standards and a staining technique that allowed for only a very small fraction of neurons to be labeled in their entirety in a sea of unlabeled neurons, Ramón y Cajal sailed uncharted anatomical waters, characterizing distinct neuronal classes based on their morphology and the architecture of their connections with other neurons. He appreciated the complex and beautiful shapes neurons adopt, and his illustrations are exquisite.[3] Ramón y Cajal surmised that the axons extending from a neuron's cell body, often for very long distances, likely conveyed information to the next neuron down the line, contacting that neuron's dendrites (arborlike processes emanating from the neuronal cell body), which in turn receive this information and propagate it to the axon and then to the next neuron's dendrites and so on. This logic allowed Ramón y Cajal to speculate about neural circuit organization throughout the nervous systems of vertebrates and even invertebrates.

In addition to adult brains from creatures of many types, Ramón y Cajal also examined embryonic nervous systems and from this work provided insight into how the complex mature nervous system is assembled. He saw that axons extending to their targets had at their tips a handlike structure we now call a growth cone, the fingers of which, called filopodia, appear to sample the external environment. When neuronal growth cones encounter a cue, either from a distance or very close by, they direct the axon toward an attractive cue and away from a repulsive cue. A wealth of data obtained over the last century has proved Ramón y Cajal extremely prescient.[4] We now know the identity of proteins secreted locally that can attract or repel extending neuronal growth cones at long distances and also of proteins associated with cell membranes that act locally to regulate neuronal growth cone guidance. We also know that axons laid down early in development can serve as scaffolds that later-developing axons follow. In this way we have begun to understand how the basic layout of complex neural connections from worms to insects to humans is elaborated. But just as a street map of New York City provides but a glimpse into its multilayered architectural and cultural heritage, we are still in the dark with respect to translating our rudimentary view of nervous system assembly into understanding how trillions of connections in the human brain are successfully wired up. Enter a useful model system: the fruit fly *Drosophila melanogaster*.

Throughout the history of biology, research organisms apparently less complex than humans have provided invaluable windows into fundamental biological processes; neuroscience is no exception. Pioneering work by several scientists, including the great geneticist Seymour Benzer, showed that *Drosophila* is an extremely useful model for studying neural development, the transmission of information across synapses from one neuron to the next, overall neural circuit organization, and even complex behaviors.[5] With its defined neuroanatomy; unmatched genetic tools; and well-characterized neuronal guidance molecules, which are remarkably similar to human neuronal guidance molecules, the fruit fly is an excellent model system to study how complex neural connections are assembled, even when these connections number far more than the available guidance cues to set them up.

The wiring of the eye to the brain in the fly is one place where we can dig into this problem of neural connections with precision. The insect compound eye consists of about 800 units, called ommatidia, that are easily visible on the surface of the eye (figure 2A). Each ommatidium includes a small lens on its outer surface (the curved "cap" you see repeated in figure 2A), and beneath each lens in the fly eye resides a group of 8 light-sensitive neurons called photoreceptors (abbreviated PR—we consider only 6 here for simplicity). Photoreceptors in an ommatidium sense light of different wavelengths, ultimately resulting in the transmission of electrical signals along their axons (figure 2B). The photoreceptor axons extend to similarly repeated units, called cartridges, in the underlying brain region, which is called the lamina. What is important is that the number and arrangement of photoreceptors within each ommatidium are invariant across all 800 or so ommatidia in each fly eye. An interesting difference among insect eyes is that for diurnal (active during the day) insects, including the butterfly, each of the photoreceptors *within* an ommatidium points in precisely the same direction in space (parallel arrows in figure 2C), and these photoreceptors from a single ommatidium extend their axons together to the same underlying cartridge (figure 2C), a relatively simple developmental event. However, insects with nocturnal activity periods, including flies such as *Drosophila*, have evolved

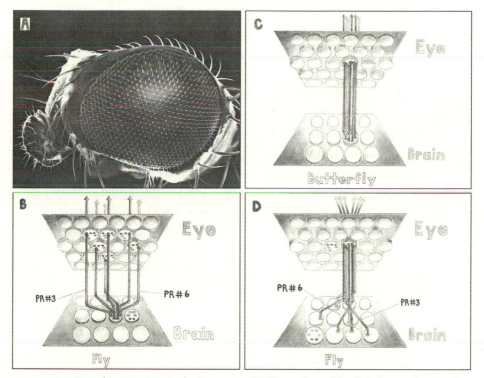

FIGURE 2.   Neural connectivity in the insect eye. (A) The surface of the fruit fly eye, showing the regular arrangement of about 800 ommatidial units. Note the curved surface over each single ommatidium; this is the lens, which focuses light onto the underlying photoreceptors. (B) A schematic showing connections between the eye and the brain in fruit fly *Drosophila*. Connections from the six different photoreceptors (PRs) in the eye that all point in exactly the same direction (arrows at top) are located in *adjacent* ommatidia and must extend axons through complex pathways to the same underlying target (cartridge) in the brain. Axons from two of these PRs are labeled PR#3 and PR#6. (C) A schematic of eye-to-brain connections in the butterfly. Unlike in the fly, PRs that point in exactly the same direction (arrows at top) are all located in a *single* ommatidium. Their axons extend directly to the same cartridge in the underlying brain, a much simpler wiring event than in the fly. (D) A schematic of eye-to-brain connections in the fly, here showing that PRs in the *same* ommatidium all point in different directions (arrows at top), and their axons extend to six different cartridges in the underlying brain.

an adaptation, called neural superposition, that increases light capture at twilight or at night without resulting in blurred vision.[6] This involves the 6 different photoreceptors, each one in an *adjacent* ommatidium, pointing precisely in the same direction (figure 2B, parallel arrows); the 6 photoreceptors in a *single* fruit fly ommatidium all point in different directions (figure 2D, divergent arrows). Yet the axons of these photoreceptors that point in the same direction and that reside in different adjacent ommatidia somehow manage to extend to the very same underlying cartridge in the lamina (figure 2B).[7] Unlike in the butterfly eye, this cannot be accomplished by simply having all photoreceptors in an ommatidium extend their axons directly down to an underlying cartridge (compare figures 2B and C), and herein lies the complexity of this wiring problem. Though figure 2B shows the wiring of only one set of 6 photoreceptors in adjacent ommatidia connecting to a single lamina cartridge in the fly brain, one must realize that this complex axon sorting is happening simultaneously for *all* the approximately 5,000 photoreceptor axons in *all* 800 ommatidia of the fly eye, a form of choreography easily outshining a Super Bowl halftime show. Producing individual labels for each photoreceptor-lamina connection is likely not the solution to preventing photoreceptor axons from forming a tangled mess as they extend across one another to their specific target cartridges.

P. Robin Hiesinger and colleagues investigated the mechanisms underlying this wiring problem in the developing fly eye in a manner Ramón y Cajal would have appreciated: they simply looked.[8] Taking advantage of modern microscopy techniques that allow imaging of individual photoreceptor axons and their growth cones as they extend to their targets, the team defined simple rules that underlie wiring in the fly eye. The microscope is a special instrument that allows the observer to peer deep inside living tissues and observe with great clarity the detailed morphology of single neurons, their axons, and their growth cones.[9] The experiment was simply to watch these photoreceptor axons and their growth cones over the approximately thirty hours it takes for them to extend from the eye into the brain during fly development. Careful analysis of these time-lapse movies allowed each of the six photoreceptors to be unambiguously identified in any one of the fly eye's 800 ommatidia.

The central observation is that each of the six photoreceptors in any individual ommatidium exhibits a different pattern of axon outgrowth once it contacts the lamina. For example, the axon from photoreceptor #3 always touches down in the lamina and then extends southeast at a fixed rate, regardless of which ommatidium in the eye one observes (figures 2B and D). The axon from photoreceptor #6 touches down in the lamina and extends west, again at a fixed rate but one different from photoreceptor #3's rate of axon extension. And so on for the other four photoreceptors. However, if one compares the axon extension patterns of photo-receptors #1–6 in *different* ommatidia across the eye, they are identical. This suggests that each of the six individual photoreceptors has a unique intrinsic growth program that is executed in the same manner in every ommatiduim, defining a rule that underlies the assembly of complex neural wiring in the fly eye. If this rule is followed and all 6 photorecep-tors extend in their designated directions and at their distinct speeds, the result is quite remarkable: the 6 photoreceptor axon growth cones extend-ing from the 6 adjacent ommatidia that see the same point in space all meet at the same time at a single lamina cartridge, and then they stop (figure 2B). This defines a second rule, which is simply that axon exten-sion ceases *only* when all 6 photoreceptor axon growth cones together contact each other and not before. So photoreceptor axons that point in the exact same direction can navigate through a teeming meshwork of axons and growth cones and still keep going since glancing contacts with one or a few growth cones that extend from photoreceptors pointing in different directions will not stop their extension. This mode of photore-ceptor wiring to the brain in the fly is extremely accurate; mistakes rarely occur, and the result is that each lamina cartridge in the brain is inner-vated only by photoreceptors that point in the exact same direction. There-fore, processing of visual stimuli at higher brain centers is greatly simpli-fied since directional information is already sorted out at the level of the lamina cartridge, the first relay station following photoreceptor sensation of light in the insect visual system. Computational modeling by Hiesinger and co-workers shows that the simultaneous meeting of the 6 photore-ceptor axon growth cones is enough to ensure correct targeting; no cue in the lamina cartridge is required. Therefore, the seemingly intractable problem of how to wire up the 5,000 photoreceptor axons, all at the same time, in the complex pattern required for neural superposition is actually accomplished by just 6 distinct photoreceptor axon growth programs. Neural superposition patterning emerges as these photoreceptor axon growth programs are executed during fly eye development. Apparently, no set of complex guidance cues is required to uniquely guide each of the approximately 5,000 photoreceptor axons to its target.

What are the implications of this work for understanding mamma-lian brain connectivity? While there is no arrangement of neurons in the human brain directly analogous to the almost crystalline organization of neurons in the fly eye, it is clear that a limited number of distinct neuro-nal cell types populate different mammalian brain regions. Neurons of the same type in the mammalian brain adopt remarkably similar pat-terns of axon and dendrite branching as they establish their unique con-nections with one another. Of course, several outside influences can act on neurons to sculpt these connections during embryonic and early post-natal neural development. These factors include guidance cues and even electrical signaling to a neuron by other neurons in a circuit. However, this work in flies reminds us that there are alternatives to instructing each individual connection in a complex neural network. It even leads to optimism regarding clinical approaches toward ameliorating damage to neurons from stroke or injury.[10] If the history of neuroscience research is any indication, we can expect this work in flies to lead to a greater un-derstanding of how simple rules establish complex connections among neurons in the human brain.[11]

NOTES

1. A combinatorial code for neural wiring is also a formal possibility. However, even if several hundred cues were able to generate a large number of unique combinations, this solution kicks the can down the road since it is an equally daunting task to precisely distribute these cues so as to selectively determine extremely complex patterns of connections among a very large number of neurons.

2. S. Ramón y Cajal, *Histology of the Nervous System*, trans. N. Swanson and L. W. Swanson (Oxford: Oxford University Press, 1995; originally published in Spanish in 1909).

3. J. DeFelipe, *Cajal's Butterflies of the Soul* (Oxford: Oxford University Press, 2010).

4. A. L. Kolodkin and M. Tessier-Lavigne, "Mechanisms and Molecules of Neuronal Wiring: A Primer," *Cold Spring Harbor Perspectives in Biology* 3 (2011): 1–14.

5. S. Benzer, "From Gene to Behavior," *Journal of the American Medical Associa-tion* 218 (1971): 1015–1022; D. Anderson and S. Brenner, "Obituary: Seymour Benzer (1921–2007)," *Nature* 451 (2008): 139.

6. E. Agi, M. Langen, S. J. Altschuler, L. F. Wu, T. Zimmermann, and P. R. Hiesinger, "The Evolution and Development of Neural Superposition," *Journal of Neurogenetics* 28 (2014): 216–232. This arrangement of neural wiring in the eyes of "advanced" flies such as *Drosophila* is called "neural superposition," and it has remained a mystery until now how this complex pattern of connec-tions between photoreceptors in adjacent ommatidia and a single underlying lamina cartridge is established.

7. Ibid.
8. M. L. Langen, E. Agi, D. J. Altschuler, L. F. Wu, S. J. Altschuler, and P. R. Hiesinger, "The Developmental Rules of Neural Superposition in Drosophila," *Cell* 162 (2015): 120–133.
9. The photoreceptor neurons were labeled using genetic tricks so that only a very few photoreceptors in any one fly eye expressed the Green Fluorescent Protein (GFP), originally isolated from the jellyfish; GFP allows individual photoreceptor neurons to be easily observed in real time as they navigate to their targets.
10. The direct introduction into the injured human brain of specific neural cell types, derived from stem cells that have been coaxed to differentiate into these cell types, is one approach under study for replacing neurons damaged by stroke or injury. It seems likely that distinct neuronal cell types have intrinsic growth programs, and an exciting possibility is that these programs might help guide the appropriate wiring of these new neurons into existing circuits and thereby facilitate repair of the nervous system.
11. Credit to Thomas Lloyd for panel A and to Natalie Hamilton for panels B–D in figure 2.

From Birth Onward, Our Experience of the World Is Dominated by the Brain's Continual Conversation with Itself

Sam Wang

A NEWBORN DOES NOT KNOW what kind of world it will encounter. What language will people speak? Will assertiveness be rewarded? What kind of food will be available? Many of a developing baby's needs arise from conditions imposed by the environment in which he or she grows up. Brains adapt to this wide variety of possibilities because developing brain circuits are strongly shaped by experience. Somehow, the baby, who at first does not have the proper connections to process a fire hose of information, gradually makes sense of the gusher.

The brain achieves this feat largely by building itself.[1] Many people think of the brain as a computational object that is programmed to make sense of incoming information and act appropriately. But contrary to the brain-as-computer metaphor, the brain does not come out of a box ready to go.[2] It takes years of experience to build a brain—and much of this construction happens well after birth. This construction process comes with enormous changes. A newborn baby's brain weighs about a pound and contains fewer than one-third the number of synaptic connections found in an adult brain. The connections it does have are mostly eliminated and replaced in the first year of life. These mostly temporary connections aren't initially organized to carry out the functions that even a two-year-old child can do, much less an adult.

Experience guides the brain's development in a highly indirect manner. Information from the outside world arrives in the brain by way of electrical impulses carried by about 15 million axons, the slender cellular wires that transmit neural messages over long distances.[3] For example, all visual information comes in through about 2 million axons going from the retina to the brain. Bodily signals of hunger, satiation, and well being come in through just 70,000 axons in the vagus nerve. And so on. That stream, which continues every moment of our lives, is passed along and further processed by tens of billions of neurons within the brain that communicate mainly with one another. In a literal sense, the brain spends most of its effort talking to itself, with the outside world a distant influence.

All this is not to say that brains are blank slates. The overall structure and wiring diagram of the brain is determined by genetic programs that work early in life, and these genetic programs also set up the principles by which neurons and synaptic connections grow and change. Experience acts within the framework of these principles. In this process, one brain region's output has a highly structured set of connections with other brain regions, and, by passing information through those connections, it can guide maturation in other parts of the brain.

To most strongly exert its influence on the developing brain, life experience has to occur during specific windows of opportunity called "sensitive periods." For vision, the sensitive period occurs in cats in the first three or four months of life and in humans for the first five to ten years of life, though the first year is especially critical. Torsten Wiesel and David Hubel discovered this principle in a series of experiments in kittens in which they found that depriving the brain of signals from one or both eyes could lead to profound disruptions in the coordination of vision between the two eyes, a process that is necessary to form a single image of the world.[4] If the brain was deprived of visual input for long enough, the disruptions to the visual parts of the brain became permanent. For example, in a procedure called "reverse eyelid suture," allowing visual images into only one eye at a time during development prevented the visual cortex from becoming organized the same way as a normal cat's; it lacked neurons that processed information from both eyes. Without these neurons, the kittens never developed normal vision (see figure 3).

Visual information from the retina arrives at the first processing station in the brain, called the thalamus. This region can extract useful fea-

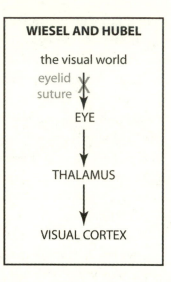

FIGURE 3. Flow of information from the visual world to the visual cortex and its interruption by eyelid suture in the classic experiments of Wiesel and Hubel.

tures and pass those on to the neocortex, much as a mother bird might prechew food before passing it on to a hungry chick. In this way, "prechewing" stages of brain processing can give other parts of the brain essential information for proper development. Wiesel and Hubel's Nobel Prize–winning work was based in part on discovering the thalamus's prechewing contribution to development. They found that to initially pioneer the path from retina to thalamus, any pattern of activity in retinal neurons would do, even that arising from diffuse light. But to refine the path from the thalamus to the visual cortex, more was required: specific patterns of activity arising from visual scenes. In the end, the ability to detect color, form, and movement requires refinements of the visual cortex that depend on having a stream of input that is passed through the thalamus. Once the thalamic input has done its "teaching" work, the thalamus continues to have the job of conveying information—no longer to an unformed circuit but rather to a sophisticated brain system for vision.

Sensitive periods arise not only in the development of vision, but also for the growth of cognitive and social abilities.[5] A devastating example occurred in Communist-era Romania when many infants and toddlers were placed in orphanages that gave them almost no tactile or social interaction. Many of these children failed to develop normal abilities in language or social communication—a syndrome that is reminiscent of autism. If the children were rescued from the orphanage by age four, they could return to a normal path of development. But if they waited too long, the changes were difficult to reverse. The brain's sensitive period for developing social abilities had passed.

The thalamus is probably not the only teacher of other brain regions. When a necessary source of information is disrupted during development, brain regions that receive the information may fail to develop properly. This idea is called "developmental diaschisis." Diaschisis (dye-AS-ki-sis; Greek *dia-*, across;-*schisis*, cut or break) is used by neurologists to describe the fact that when a brain region is damaged, activity and blood flow can change abruptly at some distant site. The probable reason is that the two brain regions are strongly connected by information-sending axons, and losing a stream of incoming information leads to sudden changes. Developmental diaschisis refers to the idea that such action-at-a-distance can have lasting and profound consequences if it happens during a developmentally sensitive period. Since many brain regions are heavily connected to one another, within-brain influences may be quite important—a form of the entire brain lifting itself up by its own bootstraps. Through an experience-guided process of brain regions getting each other organized, brains build themselves up over time (see figure 4).

My own laboratory is interested in the idea that developmental diaschisis may arise from problems in the cerebellum, which sits at the back of the brain.[6] When the cerebellum is injured in adulthood, clumsiness and uncontrolled movements result. But if the injury occurs at birth or in infancy, a very different outcome can ensue: the neurodevelopmental condition called autism spectrum disorder. Cerebellar injury at birth increases the risk of autism by a factor of forty.[7] This massive increase is on a par with the additional cancer risk that comes from cigarette smoking. Yet adults who sustain damage to the cerebellum never become autistic.

This kind of oddity is quite familiar to pediatric neurologists, who have long known that the consequences of damage to a brain region in a child can resemble the results of injuring a different brain region in adults. Such topsy-turvy clinical outcomes suggest that in babies and children, brain regions must have some kind of distant effects on one another. Autism is caused mostly by a mix of genetic and prenatal environmental factors, and one way these factors may act is by affecting the function of the cerebellum.

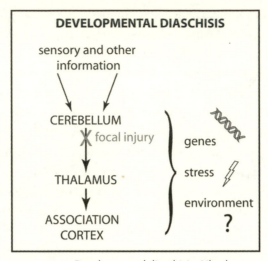

FIGURE 4. Developmental diaschisis. Like the retina of the eye, the cerebellum sends output to the thalamus, which is the principal gateway for information to pass to the cortex. Some parts of the cerebellum project to the association cortex, defined as regions that are neither sensory nor motor. During development, the interplay between the cerebellum and the cortex may be acted on by genetic programs, by stress, or by environmental events.

How does the cerebellum affect cognitive maturation? It processes many kinds of information, including sensory inputs and commands to elicit movements, within the brain to guide and refine action. It sends its output to the neocortex by way of the thalamus—the same structure whose activity is needed to guide visual development. The cerebellum is thought to predict what the world will be like one moment into the future and thus help with planning. In this way, the cerebellum may adjust and guide both movement and thought.

The developmental diaschisis hypothesis has important consequences for the treatment of autism. Developmental diaschisis opens the possibility that in early life, autism treatments may end up focusing on brain regions that were previously unsuspected to contribute to cognitive or social function, such as the cerebellum. For instance, failure of the cerebellum to predict the near future could make it hard for babies at risk for autism to learn properly from the world. Consistent with this, the most effective known treatment for autism is applied behavioral analysis, in which rewards and everyday events are paired with one another slowly and deliberately—as if compensating for a defect in some prediction process within the brain. Applied behavioral analysis works on only about half of kids with autism. It might be possible to manipulate brain activity in the cerebellum to help applied behavioral analysis work better or for more kids.

In this way, a basic principle of neuroscience may eventually help millions of children avoid the path to autism. The road to helping kids start their lifelong conversation with the rest of the world may begin with helping different parts of their brains to talk to one another.

NOTES

1. S. Aamodt and S. Wang, *Welcome to Your Child's Brain: How the Mind Grows from Conception to College* (New York: Bloomsbury, 2011).
2. A few actions are automatic at birth. For example, we're born wired to do some simple things like find mom's breast and nurse.
3. C. Cherniak, "Component Placement Optimization in the Brain," *Journal of Neuroscience* 14 (1994): 2418–2427.
4. T. N. Wiesel, "The Postnatal Development of the Visual Cortex and the Influence of Environment," Nobel lecture, December 8, 1981. http://www.nobel prize.org/nobel_prizes/medicine/laureates/1981/wiesel-lecture.html.
5. E. I. Knudsen, "Sensitive Periods in the Development of the Brain and Behavior," *Journal of Cognitive Neuroscience* 16 (2004): 1412–1425.
6. S. S-H. Wang, A. D. Kloth, and A. Badura, "The Cerebellum, Sensitive Periods, and Autism," *Neuron* 83 (2014): 518–532.
7. S. Wang, "How to Think about the Risk of Autism," *New York Times*, March 29, 2014, p. SR6.

Children's Brains Are Different

Amy Bastian

I RECENTLY WENT SKIING for the first time in many years. I was out of shape and definitely out of practice, so I began by carefully working my way down one of the easier slopes. Suddenly, out of nowhere, a tiny child zipped right in front of me. I lurched out of the way and fell, losing a ski and a pole in the process. The child looked like he was barely old enough to walk, let alone ski! I was initially annoyed—where was this kid's parent? Should he be skiing alone like that? After my frustration subsided, I made some effort to watch tiny children rocket down steep hills. How on earth do young children learn to ski so well?

Very young children appear to learn many skills better than adults, in such areas as sports, languages, and musical instruments. In fact, most coaches and teachers will tell you that you must start young if you are going to become, for example, a great tennis player or a concert violinist. And you typically have to start young if you want to learn a second language as proficiently as a native speaker. So why is this the case— what is special about a child's brain? Are young children always super-learners compared to adults? Even more important, is there any downside to superior learning abilities in childhood? I will address each of these questions in turn. To be completely honest, the short answer is that we don't *really* understand the how's and why's of learning abilities in human development. But we do know a few interesting things.

What is special about a child's brain? Ask almost any neuroscientist, and he or she will probably say, "A child's brain is more plastic." This answer is not terribly helpful because it does not address what "plastic" means, what makes a brain plastic, or why it becomes less plastic in adulthood. Here I will use a simple definition of "plasticity": the ability of the brain to change its own connections and functions as a result of new experiences. There are many cellular and network level mechanisms in the brain that contribute to this plasticity (see the essay by Linda Wilbrecht in this volume).[1]

One of the most dramatic aspects of brain development that could underlie plasticity occurs in infancy and early childhood and involves a massive proliferation of neural connections.[2] The brain of a two-year-old child has twice as many neural connections as the adult brain. The number of contacts between neurons (i.e., synapses) explodes during infancy, with some estimates suggesting that hundreds of new synapses are formed every second! This is a highly dynamic process—connections are changing constantly throughout early life. There are chemical signals in the developing brain that help guide the correct connections and repel incorrect ones. This abundance of neural connections is eventually pruned back throughout childhood and adolescence to reach adult levels.

One important influence on whether connections stay or go is whether those connections are used. Thus the variety, intensity, and type of experiences of infants and young children are incredibly important for their developing brains.[3] Connections that are used as a child moves, listens, sees, thinks, and feels are the ones that are more likely to stick. Without doing these things, the connections may be weakened or removed. Thus the child's brain structure may be optimized early on for learning very different kinds of things, ranging from speaking Mandarin Chinese to playing professional tennis. The important step is that the child engage in these activities so that the right connections are laid down. This is, of course, a crude simplification of the amazing and complex processes that are going on within young brains. But experience-dependent brain plasticity in childhood is undoubtedly an important factor that may produce a heightened learning ability in children for specific behaviors.

It is also important to define what constitutes heightened learning ability in children compared to adults. We think of children as super-learners across the board, but is this really the case? It depends on what we mean by "super-learner." Learning can be measured many different ways—for example, how fast you learn, how much you learn, the quality of what you learn, and how much you retain. And there are many types of learning that depend on different brain systems and are driven by unique behaviors, so learning in one domain may not transfer to another. Consider learning a second language. Children are super-learners in the sense that they can learn to be more proficient compared to adults—that is, they can gain fluency in a second language that is comparable to that of a native speaker. But this does not mean that every aspect of language learning is better. In fact, children learn a second language more slowly than adults; it takes them longer to learn to read, speak words, and use the appropriate grammatical rules.[4] So young children are ultimately better in proficiency but not in their speed of language acquisition.

Similarly, it appears that younger children learn new movements at a slower rate compared to adults. Some work in this area has shown that this motor learning rate gradually improves (i.e., speeds up) through childhood and becomes adultlike by about age twelve.[5] Children also start at a lower level of motor proficiency compared to adults; they are more variable and less accurate in their movement.[6] The lower proficiency is likely because parts of the brain that are involved in movement control are still maturing throughout childhood.

If children learn more slowly and are more variable in their movements, why do they appear to learn certain tasks like skiing better than adults? First, children are smaller than adults; thus their center of mass is lower, a factor that may make activities like skiing easier to learn to control. However, this factor would not explain their learned proficiency across fine motor skills, such as playing a video game, an activity that involves only hand movements. Second, the variability in children's movements might also work to their advantage, as they get to try out many different ways of moving for a given situation in order to find the best one. We know that this movement exploration is an essential part of motor learning. Adults may be less willing to explore different movements and therefore tend to settle on a suboptimal motor pattern. Third and perhaps the most important factor is that children may be more willing than adults to undergo massive amounts of practice to learn motor skills. For example, when infants are learning to walk, they take about 2,400 steps and fall seventeen times for *each hour* of practice. This is an intense amount of activity; it means that infants cover the length of about 7 American football fields per hour! And in the six hours of the day that they might be active, they will fall a hundred times and travel the length of forty-six football fields.[7] Thus the intensity of practice that infants and children are willing to undergo, coupled with the heightened level of experience-dependent plasticity in the child's brain, may be why they can learn motor skills to levels beyond those of adults.

Unfortunately, there is also a downside to experience-dependent plasticity in childhood because *any kinds* of experiences affect brain development, not just the positive ones. So although plasticity can make children learn better, it can also cause problems. Stress and negative experiences can lead to maladaptive changes in a child's brain.[8] For example, young children who experience events such as neglect, abuse, or poverty have an increased risk of developing problems such as anxiety, emotional dysfunction, and cognitive deficits. It is thought that these problems are not merely a direct reaction to the negative experiences but also reflect a fundamental change in the brain circuitry that mediates these processes. Further, the lack of experience can have extremely deleterious effects during development.[9] If a young child has to have one eye patched for an extended period of time, thereby occluding vision, it can lead to irreversible changes in the development of visual areas in the brain and difficulties with depth perception. Similarly, young children who are not read to when they are young can show slowed learning of language and poorer literacy.[10]

Ultimately, all experiences, as well as the lack of them, count a lot during early development. Children can take advantage of early plasticity to learn many things better than adults, including zipping down a ski slope and speaking French. But this plasticity can also put children at risk when they have negative experiences or are deprived early in life. Scientists don't fully understand the processes that contribute to childhood brain plasticity, but it seems clear that early life experiences are incredibly important. Future work will help to uncover how we can make the most of this remarkable time of life.

NOTES

1. J. Stiles and T. L. Jernigan, "The Basics of Brain Development," *Neuropsychology Review* 20 (2010): 327–348.

2. G. M. Innocenti and D. J. Price, "Exuberance in the Development of Cortical Networks," *Nature Reviews Neuroscience* 6 (2005): 955–965.

3. W. T. Greenough, J. E. Black, et al., "Experience and Brain Development," *Child Development* 58 (1987): 539–559.

4. J. S. Johnson and E. L. Newport, "Critical Period Effects in Second Language Learning: The Influence of Maturational State on the Acquisition of English as a Second Language," *Cognitive Psychology* 21 (1989): 60–99.

5. E. V. Vasudevan, G. Torres-Oviedo, S. M. Morton, J. F. Yang, and A. J. Bastian, "Younger Is Not Always Better: Development of Locomotor Adaptation from Childhood to Adulthood," *Journal of Neuroscience* 31 (2011): 3055–3065.

6. Ibid.; R. Gómez-Moya, R. Díaz, and J. Fernandez-Ruiz, "Different Visuomotor Processes Maturation Rates in Children Support Dual Visuomotor Learning Systems," *Human Movement Science* 46 (2016): 221–228.

7. K. E. Adolph, W. G. Cole, M. Komati, J. S. Garciaguirre, D. Badaly, J. M. Lingeman, G. L. Chan, and R. B. Sotsky, "How Do You Learn to Walk? Thousands of Steps and Dozens of Falls per Day," *Psychological Science* 23 (2012): 1387–1394.

8. G. Turecki, V. K. Ota, S. I. Belangero, A. Jackowski, and J. Kaufman, "Early Life Adversity, Genomic Plasticity, and Psychopathology," *Lancet Psychiatry* 1 (2014): 461–466.

9. D. H. Hubel and T. N. Wiesel, "The Period of Susceptibility to the Physiological Effects of Unilateral Eye Closure in Kittens," *Journal of Physiology* 206 (1970): 419–436.

10. P. C. High, L. LaGasse, S. Becker, I. Ahlgren, and A. Gardner, "Literacy Promotion in Primary Care Pediatrics: Can We Make a Difference?" *Pediatrics* 105 (2000): 927–934.

Your Twelve-Year-Old Isn't Just Sprouting New Hair but Is Also Forming (and Being Formed by) New Neural Connections

Linda Wilbrecht

SAYING THE PHRASE "the teenage brain" can, by itself, elicit a knowing chuckle. People who might otherwise claim to know nothing about brains can deliver a lecture on "the teenage brain." Such brains are wild; crazy; full of raging hormones. Their frontal lobes are under construction or are just coming "online." Teenagers are out of control. They have no brakes. But we have to be careful that this pop-culture view of teenagers doesn't distort how we study and understand them. What is actually going on in their brains as they move through the teenage years? If you take a close look at what their neurons are up to, you might change your parenting style or, at least, roll your eyes less when someone mentions teenagers.

When we look at teenage brains in a scanning machine, they do look different from those of both children and adults. The neural activity occurs to different extents and in different places.[1] Over the teenage years, the gray matter is thinning and the white matter is growing.[2] It can take until the midtwenties until brain scans look adultlike, with the frontal lobes, a region involved in self-control, planning, and foreseeing consequences of one's actions, as the last area to mature.[3] That the frontal cortex does not yet appear adultlike is marched out in practically any issue concerning teens: screen time, drug use, voting, sexual behavior.[4] The "immature" frontal cortex is blamed for the acts teenagers do that we do

45

not like, and it is used to justify why teenagers should be prevented from gaining access to things that are dangerous or powerful.

It is easy to focus on the negative and view teenagers as transiently deranged by their biology or in a state comparable to frontal lobotomy.[5] Yet if you take a closer look at what is going on inside the brain, you might warm with the pride of a grandmother. There is no lobotomy to be found, no black hole where the frontal lobes should be. There are neurons there, and they are up to something that looks pretty creative, smart, and useful.

In the last two decades, new imaging technology has granted us a view into what is happening to individual neurons in living tissue in mice and other laboratory animals.[6] Previously, we could look at what the neurons were doing in snapshots taken from post-mortem tissue fixed at a particular moment in time. Since around the year 2000, we have been able to use special laser scanning microscopes to follow the neurons in the mouse brain as they grow up and before and after the brain has a new experience. In terms of getting to know what the neurons are doing and what they are like, this is like going from having a single black-and-white photo to having hours of childhood video.

We can now see that juvenile and adolescent frontal lobe neurons are busy *exploring*. They are hungrily exploring all they can know in the world, and such exploration mainly concerns their potential connection to other neurons in the brain.[7]

Neurons are shaped like craggy trees and bushes. Even before we hit puberty, the neurons have already achieved their full height, and their branches and roots are tightly interwoven in a dense thicket. In lab animals like mice, we can illuminate a single neuron at a time inside this thicket and take pictures or videos of it as it matures. What we see in the late childhood and adolescent brain are a multitude of changes in tiny, thornlike structures called dendritic spines (see figure 5). From looking at still images of dead tissue, we know that these dendritic spines become less numerous by the time humans and common lab animals reach the early adulthood phase.[8] When, however, a neuron is alive, we can see these spines are sprouting too, extending and retracting as they explore the beckoning outputs of other neurons.[9] Information can pass between neurons when one spine firmly connects to the arbor of another neuron by making a synaptic connection at the end of the spine. This connection

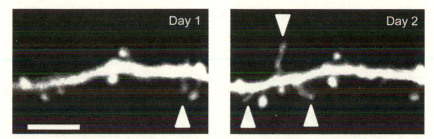

FIGURE 5. Time-lapse image of new spines sprouting overnight in an adolescent mouse. Arrows indicate a spine that will be lost after day 1 and new spines that are gained between day 1 and day 2. Scale bar = 5 microns. Photo credit: Josiah Boivin (Wilbrecht lab).

can later be broken when the spine retracts back into the dendritic branch from which it originated.

Repeatedly observing neurons, day after day, we can see them sprouting these new connections and then losing most of them again on subsequent days. In these new growths and retractions, we presume they are exploring their potential connectivity to other neurons. As the brain nears puberty, neurons may grow and lose 25 percent or more of their connections every week.[10] As the brain enters young adulthood, this turnover of connections can sink to 10 percent or less, depending on the brain region examined. Because the connectivity of a neuron is important for its function in the network, the functional identity of each neuron can undergo radical change from week to week in the developing brain. (Just imagine you were 25 percent different next week! What might your family say?) As we become adults, the total number of connections diminishes, and the potential for a new connection wanes.

What useful conclusions might be drawn from knowledge of neuronal exploration in the teenage frontal lobes? The large-scale turnover of neural connections might explain why the frontal lobe may not be as efficient in teenagers as compared to adults. However, it may also enable greater capacity for some forms of learning or flexibility in the face of change. These neural connections in the frontal lobes may be the main substrate upon which an individual's adult personality and tendencies are built. We can imagine that a unique, individual mind may readily be sculpted from this rapidly budding neural topiary. This raises the following questions: How is this topiary shaped? And by whom? Or by what?

This is where experience comes into the picture. As you read this, in each developing brain across the world, unfathomable numbers of new synapses are in a tenuous state. Which will survive and why? As best we can tell, trial-and-error learning gained through active experience drives this process. By observing and quantifying changes in the gains and losses in connections between neurons, researchers can see that a whole cohort of new connections may be kept when a new skill or rule is learned. For example, the motor areas stabilize a crop of newly sprouted spines as a new motor skill is learned.[11] New frontal lobe connections are also sustained when the brain is learning that two things tend to go together— for example, when a sound, sight, or smell is associated with something painful[12] or pleasurable.[13]

Most recently, experiments have suggested that neurons also track aspects of the self. That is, growth and pruning of synaptic connections in the frontal lobe seems to be doing more than just reflecting what happened in the external world and whether it was good or bad. This process also appears to be tracking self-generated strategy along with outcome: "What did I just try in the world?" and "Was that good or bad for me?"[14] These findings suggest that self-generated trial-and-error exploration is playing a role in the formative shaping of the frontal neural topiary. So you might differentially sculpt your frontal circuits when actively doing something versus just passively observing.[15]

If you are not fond of gardening, topiary, teenagers, or neurons, you might question whether these observations are important. You might think, "So what! This has been happening quietly in all the frontal lobes in the long history of mammals; it changes nothing now that we can see it. The bottom line is still that the frontal lobes in teenagers are immature." However, I think seeing and imagining the sprouting, connecting neurons undergoing their own formative years turns things 180 degrees.

If we go back to the beginning and imagine that teenagers are comparable to patients lacking a frontal lobe, then we might decide that they need protection from themselves and from the world. We might place them away in a safe space and just wait for them to grow up.[16] On the other hand, if we imagine that the frontal lobes are populated by neurons wildly grasping for information upon which to form themselves and we realize their capacity to change is waning by the day, then we will want to thrust teenagers out into the world of harsh life lessons. Sign them up for Arctic wilderness camp!

Of course, these strategies are extreme ends of the spectrum, but the plight of countless new neural connections reframes teenage experience from something frivolous to something serious. It suggests warehousing teens in overwhelmed schools, refugee camps, and the like is a shameful waste of opportunity from which a generation may never recover. It suggests that pretend play is a necessary precursor to work. It suggests that some painful falls might actually be helpful for growing up.[17] Of course, we should not throw out centuries of wisdom on how and when to protect teenagers. We probably should not allow them unfettered access to drugs and video games. They do need a guide, they do need scaffolding, but their neurons are alive, exploring, and taking their shape from these experiences. These experiences are, in turn, likely to shape a person for life.

NOTES

1. S. Durston, M. C. Davidson, N. Tottenham, A. Galvan, J. Spicer, J. A. Fossella, and B. J. Casey, "A Shift from Diffuse to Focal Cortical Activity with Development," *Developmental Science* 9 (2006): 1–8.

2. B. J. Casey, A. Galvan, and T. A. Hare, "Changes in Cerebral Functional Organization during Cognitive Development," *Current Opinion in Neurobiology* 15 (2005): 239–244.

3. N. Gogtay, J. N. Giedd, L. Lusk, K. M. Hayashi, D. Greenstein, A. C. Vaituzis, T. F. Nugent, D. H. Herman, L. S. Clasen, A. W. Toga, and J. L. Rapoport, "Dynamic Mapping of Human Cortical Development during Childhood through Early Adulthood," *Proceedings of the National Academy of Sciences of the USA* 101 (2004): 8174–8179.

4. The immature frontal lobe construct is usually included in most news stories about teenagers and risky behavior. Here are some examples: http://www.cnn.com/2014/07/15/health/science—drinking—age/; http://www.vice.com/read/how-screen-addiction-is-ruining-the-brains-of-children.

5. Tests that were used to identify deficits in frontal lobe function after brain damage were subsequently administered to kids and teenagers to determine when the frontal lobes become adultlike. It is interesting that kids look pretty adultlike in many of these "frontal tests" by the time they are ten to twelve years old, although some aspects improve through the period from twelve to twenty-five. G. J. Chelune and R. A. Baer, "Developmental Norms for the Wisconsin Card Sorting Test," *Journal of Clinical and Experimental Neuropsychology* 8 (1986): 219–228; H. S. Levin, K. A. Culhane, J. Hartmann,

K. Evankovich, A. J. Mattson, H. Harward, G. Ringholz, L. Ewing-Cobbs, and J. M. Fletcher, "Developmental Changes in Performance on Tests of Purported Frontal Lobe Functioning," *Developmental Neuropsychology* 7 (1991): 377–395; M. Huizinga, C. V. Dolan, and M. W. van der Molen, "Age-Related Change in Executive Function: Developmental Trends and a Latent Variable Analysis," *Neuropsychologia* 44 (2006): 2017–2036.

6. W. Denk and K. Svoboda, "Photon Upmanship: Why Multiphoton Imaging Is More Than a Gimmick," *Neuron* 18 (1997): 351–357.

7. A. Holtmaat and K. Svoboda, "Experience-Dependent Structural Synaptic Plasticity in the Mammalian Brain," *Nature Reviews Neuroscience* 10 (2009): 647–658; A. Stepanyants and D. B. Chklovskii, "Neurogeometry and Potential Synaptic Connectivity," *Trends in Neurosciences* 28 (2005): 387–394.

8. P. R. Huttenlocher, "Synaptic Density in Human Frontal Cortex—Developmental Changes and Effects of Aging," *Brain Research* 163 (1979): 195–205; Z. Petanjek, M. Judaš, G. Šimić, M. R. Rašin, H. B. Uylings, P. Rakić, and I. Kostović, "Extraordinary Neoteny of Synaptic Spines in the Human Prefrontal Cortex," *Proceedings of the National Academy of Sciences of the USA* 108 (2011): 13281–13286.

9. G. W. Knott, A. Holtmaat, L. Wilbrecht, E. Welker, and K. Svoboda, "Spine Growth Precedes Synapse Formation in the Adult Neocortex in vivo," *Nature Neuroscience* 9 (2006): 1117–1124; C. C. Chen, J. Lu, and Y. Zuo, "Spatiotemporal Dynamics of Dendritic Spines in the Living Brain," *Frontiers in Neuroanatomy* 8 (2014): 28.

10. Y. Zuo, A. Lin, P. Chang, and W. B. Gan, "Development of Long-Term Dendritic Spine Stability in Diverse Regions of Cerebral Cortex," *Neuron* 46 (2005): 181–189; A. J. Holtmaat, J. T. Trachtenberg, L. Wilbrecht, G. M. Shepherd, X. Zhang, G. W. Knott, and K. Svoboda, "Transient and Persistent Dendritic Spines in the Neocortex in vivo," *Neuron* 45 (2005): 279–291; C. M.Johnson, F. A. Loucks, H. Peckler, A. W. Thomas, P. H. Janak, and L. Wilbrecht, "Long-Range Orbitofrontal and Amygdala Axons Show Divergent Patterns of Maturation in the Frontal Cortex across Adolescence," *Developmental Cognitive Neuroscience* 18 (2016): 113–120.

11. T. Xu, X. Yu, A. J. Perlik, W. F. Tobin, J. A. Zweig, K. Tennant, T. Jones, and Y. Zuo, "Rapid Formation and Selective Stabilization of Synapses for Enduring Motor Memories," *Nature* 462 (2009): 915–919.

12. C. S. W. Lai, T. F. Franke, and W. B. Gan, "Opposite Effects of Fear Conditioning and Extinction on Dendritic Spine Remodelling," *Nature* 483 (2012): 87–91.

13. F. J. Muñoz-Cuevas, J. Athilingam, D. Piscopo, and L. Wilbrecht, "Cocaine-Induced Structural Plasticity in Frontal Cortex Correlates with Conditioned Place Preference," *Nature Neuroscience* 16 (2013): 1367–1369.

14. C. M. Johnson, H. Peckler, L. H. Tai, and L. Wilbrecht, "Rule Learning Enhances Structural Plasticity of Long-Range Axons in Frontal Cortex," *Nature Communications* 7 (2016): 10785

15. This is probably no surprise to a teacher but hopefully good ammunition to support the extra effort to create active learning environments.
16. Chris Erskine of the *Los Angeles Times* pretends to cope with pubertal changes in his thirteen-year-old by locking him in a puppy crate but then sends him to camp instead; http://www.latimes.com/home/la-hm-erskine-20160718 -snap-story.html.
17. This idea is also represented in popular parenting books like Wendy Mogel's *The Blessing of a Skinned Knee* (New York: Scribner, 2001) and Jessica Lahey's *The Gift of Failure* (New York: Harper Collins, 2015).

How You Use Your Brain Can Change Its Basic Structural Organization

Melissa Lau and Hollis Cline

CERTAIN MEMORIES STICK WITH YOU—for example, sitting in the sunshine at your graduation. Or that embarrassing piano recital when you were eight. Or the first time you held your child. Or that time a bat flew into your house. Each single experience can leave a biological trace because memories are formed, at least in part, by changing the connections between neurons. But how is your brain affected by years of repetitive training on a single subject? How do those incremental changes add up? Is it possible to see dramatic alterations to the fundamental organization of the brain from repeated experience?

For London taxi drivers, an intimate knowledge of the city's twenty-five thousand streets and twenty thousand landmarks is hard won. To earn their license, the drivers are required to recite the shortest route between any two points in that chaotic city. Even after several years of studying, not everyone passes the exams to become a taxi driver. Is this remarkable navigational skill reflected by measurable differences in their brains? In fact, when compared to the general public, London taxi drivers have larger posterior hippocampi—a region involved in spatial memory.[1] But does their rigorous training actually change their brains, or are people with naturally large hippocampi just more likely to pass the taxi driver exams? How much effect, if any, can our individual experiences have in shaping our brains?

Like humans, birds use their hippocampi for spatial maps and memory. Unlike taxi drivers, some bird species have seasonal fluctuations in the size of this brain region. For example, black-capped chickadees have larger hippocampi in October, which also happens to be the peak season for food-hoarding.[2] These birds stash their food in multiple locations for later meals. Given the dramatic seasonal differences in hippocampal volume (it's 30 percent larger in October than in August!), it's tempting to speculate that this brain region enlarges because the chickadees need to remember where their food is hidden.

Other seasonal behaviors in birds have also been linked to changes in the brain. The size of the brain region called HVC, which is involved in song production, fluctuates for the male birds of some species.[3] Male great tits, who sing complex courtship and territorial songs during the breeding season, have larger HVCs in the spring. In contrast, willow tits, who sing all year round, have no seasonal changes in HVC volume.[4] However, it's not clear what causes the seasonal changes in either of these brain regions. Such changes in volume could be driven by an environmental trigger (like temperature or length of day) to prepare for seasonal behaviors like food hoarding or singing. Alternatively, is it possible that certain brain regions can expand from increased use?

To address that question, several research groups began training monkeys in a variety of tasks. For example, adult owl monkeys were trained to touch a rotating disk.[5] The whole contraption was placed just within reach so that if the monkey held its fingertips on it, he'd be rewarded with a banana-flavored pellet. Much like a spinning record (but with raised bumps on it), this machine delivered a steady stream of tactile stimulation to the monkey's fingertips. By mapping the monkey's brain activity before and after training, the researchers tested whether repeated stimulation of the fingertips induced changes in the somatosensory cortex, the part of the brain that processes touch. In just a few months, there were already measurable differences.

The somatosensory cortex can be divided into separate regions, each dedicated to different body parts. After the training, more of the somatosensory cortex was devoted to processing touch in the fingers—and specifically only in those fingertips that were stimulated. Because adult brains no longer generate new neurons (except for certain regions, like the hippocampus) the size of the cortex is fixed and becomes valuable real estate.

Like squabbling landowners competing to expand the borders of their prescribed properties, increased cortical representation of the fingertips comes at the expense of neighboring brain regions. In this case, more of the cortex is assigned to the stimulated fingertips; that increase comes from a loss in representation of neighboring (unstimulated) fingers and even a shift in the border between hand and face. Here, experience, or the use of specific neural circuits, does expand the cortical area allotted to that brain function—but it is not without costs.

This strategy of compromise, where increased cortical representation of a single function comes at the loss of another, is a general principle that's been observed in a variety of scenarios. Like the somatosensory cortex, the motor cortex is organized into a map where portions of the cortex are responsible for directing movement in specific body parts. By training squirrel monkeys to complete tasks that isolate certain muscle groups, researchers can look for corresponding changes in those specific parts of the motor cortex. In a task that requires skilled finger movements, monkeys are trained to remove banana-flavored pellets from a tiny hole. Alternatively, other monkeys are trained in a key-turning task that requires forearm motion. Repeated use of the fingers increased the area of motor cortex devoted to finger movement, and the increase came at the expense of the neighboring area for the forearm. Likewise, repeated training for key-turning increased representation of the forearm area while decreasing the cortical space for fingers.[6] As for the permanency of these changes, the phrase "use it or lose it" comes to mind. After the monkeys stopped training, the motor cortex shifted back toward the original representation of these different body parts.

In humans, specific types of training can also lead to discrete changes in the organization of the somatosensory and motor cortices. Much like the monkeys touching the rotating disk, blind people who learn Braille have measurable differences in their sensorimotor cortex.[7] Their reading finger has a larger cortical representation than their nonreading fingers and is larger than the corresponding representation for fingers of non-Braille readers.

Musicians also have significant differences in their motor cortices compared to nonmusicians.[8] Those who play stringed instruments, such as the violin or guitar, become extremely skilled in finger movements of the left hand. The right hand, which carries a bow or strums the strings, generally requires less dexterity. Strikingly, these string players have a larger cortical representation for the left hand than do nonmusicians. The part of motor cortex responsible for the right hand wasn't different between musicians and nonmusicians. Furthermore, musicians who had been practicing longer had more cortical reorganization than novice musicians. Although there was a correlation between amount of training and extent of brain changes, such a correlation doesn't definitively prove that musical training causes changes in the motor cortex. For example, what if people with larger left-hand cortical representations happen to be better string musicians and so are more likely to continue playing?

How can a scientist test whether training induces cortical changes? He or she can teach people how to juggle. In this clever, beautifully simple experiment, people's brains were scanned before and after learning to juggle.[9] After three months of practice, the volunteers could juggle for an entire minute without mistakes—and there were distinct changes in their brains. Structural magnetic resonance imaging (MRI),[10] used here to examine the anatomical structure of the brain, revealed a selective expansion in the gray matter of the mid-temporal area, the part of the brain that processes the speed and direction of moving objects. There was also an enlargement of the brain region for perceptual motor coordination and visual attention, all components of the skills needed to become a proficient juggler. Three months later, after a break from practicing, most volunteers could no longer juggle, and the corresponding brain expansions had reversed. In just six months, this experiment showed that training causes transient, but very real, structural changes in the brain! Unlike the changes in functional representation seen in the somatosensory and motor cortices, this type of training actually *increases* the volume of certain brain areas.[11] The juggling study didn't explore direct mechanisms for brain expansion, but it's possible that specific brain regions can grow— even in the absence of newly born neurons—if individual cells become larger. For instance, alterations in cell size might include both a larger network of neuronal processes and an increased number of synapses. For sure, it appears that training can affect structure and function of the brain through a myriad of ways.

Unfortunately, the incredible plasticity of the brain can, on occasion, have unwanted consequences. For example, many amputees experience a "phantom limb," the sensation that their limb is still attached. Some even have pain in the phantom limb. One leading theory is that cortical reorganization is responsible for phantom pain. For example, the parts of the somatosensory and motor cortex that were originally allotted to an amputated hand are eventually taken over by neighboring regions— namely, the lips.[12] Some data suggest that this functional invasion of the lower face region into the hand region (and thus the subsequent activity) is responsible for phantom pain.[13] However, while phantom pain is quite common, only some amputees feel sensation in their phantom limb when touched elsewhere on their bodies.[14] This observation suggests that additional mechanisms are at work. Other studies indicate that residual connections from the missing hand activate phantom pain,[15] and changes in the excitability of the spinal cord may contribute as well.[16] Regardless, there are many examples in which increased cortical representation of one area comes at the expense of a neighboring area.

Might there be functional consequences for this reorganization of limited cortical resources? Let's return to the London taxi drivers. What we have yet to mention is that the taxi drivers' expansion of the posterior hippocampus comes at the cost of the anterior hippocampus.[17] The overall volume of the hippocampus is the same between drivers and controls; it's just the regional volumes that differ. The posterior hippocampus is thought to store spatial representation of the environment, such that an expansion here could allow for a more detailed mental map. In contrast, the corresponding reduction in anterior hippocampus might explain some of the functional deficits seen in taxi drivers. Most broadly, they're worse than nondrivers at forming new visual and spatial memories. For example, when given a complex line drawing to copy, they're worse at redrawing the figure in a later memory test; this task tests the ability to remember how visual elements are spatially arranged.[18]

A more recent study, following a group of prospective taxi drivers over the course of four years, was able to definitively show that training causes changes in the hippocampus. After several years of studying, the trainees that passed the taxi drivers' exams had an expanded posterior hippocampus and performed worse on visual and spatial tasks. In contrast, the brains of trainees who failed or dropped out were no different from controls. It's the experience of training itself that drives structural changes in the hippocampus, and it can have unintended functional consequences, like the deficit in forming visual and spatial memories.

Still, it's important to remember that this isn't necessarily bad—it's just your brain's response to meet the functional demands of your environment. Just look at retired London taxi drivers. They have smaller posterior hippocampi and better visual and spatial memory than full-time drivers.[19] With decreased demands on the neural circuitry for spatial navigation, the brain seems to shift back to conditions seen in nondriver controls. This all illustrates that our brains are constantly changing.

Amazingly, it's our daily experiences that can alter the basic organization of our brains in dramatic, tangible ways. The utility of this biological phenomenon is elegant in its simplicity: it is the brain that defines our perception of the world around us, and yet, with beautiful symmetry, it is our perceptual experiences that can shape the underlying structure and functional capabilities of the brain itself.

NOTES

1. E. A. Maguire et al., "Navigation-Related Structural Change in the Hippocampi of Taxi Drivers," *Proceedings of the National Academy of the Sciences of the USA* 97 (2000): 4398–4403; E. A. Maguire, K. Woollett, and H. J. Spiers, "London Taxi Drivers and Bus Drivers: A Structural MRI and Neuropsychological Analysis," *Hippocampus* 16 (2006): 1091–1101; K. Woollett and E. A. Maguire, "Navigational Expertise May Compromise Anterograde Associative Memory," *Neuropsychologia* 47 (2009): 1088–1095; K. Woollett, H. J. Spiers, and E. A. Maguire, "Talent in the Taxi: A Model System for Exploring Expertise," *Philosophical Transactions of the Royal Society of London B: Biological Sciences* 364 (2009): 1407–1416; K. Woollett and E. A. Maguire, "Acquiring 'the Knowledge' of London's Layout Drives Structural Brain Changes," *Current Biology* 21 (2011): 2109–2114.

2. T. V. Smulders, A. D. Sasson, and T. J. DeVoogd, "Seasonal Variation in Hippocampal Volume in a Food-Storing Bird, the Black-Capped Chickadee," *Journal of Neurobiology* 27 (1995): 15–25.

3. While the birds discussed here and studied in Longmoor et al. (see note 4 below) were all males, other research suggests that seasonal changes to song system brain structures occur in female birds of other species. S. A. MacDougall-Shackleton et al., "Photostimulation Induces Rapid Growth of Song-Control Brain Regions in Male and Female Chickadees (*Poecile atricapilla*)," *Neuroscience Letters* 340 (2003): 165–168. Overall though, female song systems haven't been studied as much as in males—especially since it has only recently been accepted that female song is actually prevalent in many songbird species. K. J. Odom et al., "Female Song Is Widespread and Ancestral in Songbirds," *Nature Communications* 5 (2014): 3379.

4. G. K. Longmoor et al., "Different Seasonal Patterns in Song System Volume

in Willow Tits and Great Tits, *Brain Behavior and Evolution* 87 (2016): 265–274.

5. W. M. Jenkins et al., "Functional Reorganization of Primary Somatosensory Cortex in Adult Owl Monkeys after Behaviorally Controlled Tactile Stimulation," *Journal of Neurophysiology* 63 (1990): 82–104.

6. R. J. Nudo et al., "Use-Dependent Alterations of Movement Representations in Primary Motor Cortex of Adult Squirrel Monkeys," *Journal of Neuroscience* 16 (1996): 785–807.

7. A. Pascual-Leone and F. Torres, "Plasticity of the Sensorimotor Cortex Representation of the Reading Finger in Braille Readers," *Brain* 116 (1993): 39–52.

8. T. Elbert, C. Pantev, C. Wienbruch, B. Rockstroh, and E. Taub, "Increased Cortical Representation of the Fingers of the Left Hand in String Players," *Science* 270 (1995): 305–307.

9. B. Draganski et al., "Neuroplasticity: Changes in Grey Matter Induced by Training," *Nature* 427 (2004): 311–312.

10. Not to be confused with another common neuroimaging technique, functional MRI (fMRI), which is used to infer regions of brain activity based on dynamic changes in blood oxygen level.

11. Note that the previous studies (Jenkins et al. [note 5]; Nudo et al. [note 6]) didn't use techniques that specifically measure structural changes. Instead, they focused on testing the functional consequences of sensory and motor experiences. There may well have been (unrecorded) changes in volume that accompanied the observed functional reorganization. In fact, training-induced structural changes *have* been recorded in the somatosensory cortex—namely, increased cortical thickness in the lip region of wind instrument players. U. S. Choi et al., "Structural and Functional Plasticity Specific to Musical Training with Wind Instruments," *Frontiers in Human Neuroscience* 9 (2015): 597. It's possible that structural and functional changes are concurrent. But to know for sure, each training scenario has to be tested with both structural and functional analyses. For example, some training regimens induce functional reorganization without overall changes in brain volume (Maguire et al. [note 1 above]).

12. T. Elbert et al., "Input-Increase and Input-Decrease Types of Cortical Reorganization after Upper Extremity Amputation in Humans," *Experimental Brain Research* 117 (1997): 161–164; H. Flor et al., "Phantom-Limb Pain As a Perceptual Correlate of Cortical Reorganization Following Arm Amputation," *Nature* 375 (1995): 482–484; E. Raffin et al., "Primary Motor Cortex Changes after Amputation Correlate with Phantom Limb Pain and the Ability to Move the Phantom Limb," *Neuroimage* 130 (2016): 134–144.

13. Flor et al., "Phantom-Limb Pain As a Perceptual Correlate of Cortical Reorganization Following Arm Amputation"; H. Flor, L. Nikolajsen, and T. Staehelin Jensen, "Phantom Limb Pain: A Case of Maladaptive CNS Plasticity?" *Nature Reviews Neuroscience* 7 (2006): 873–881.

14. Flor, Nikolajsen, and Staehelin Jensen, "Phantom Limb Pain: A Case of

Maladaptive CNS Plasticity?"; S. M. Grusser et al., "The Relationship of Perceptual Phenomena and Cortical Reorganization in Upper Extremity Amputees," *Neuroscience* 102 (2001): 263–272.

15. T. R. Makin et al., "Phantom Pain Is Associated with Preserved Structure and Function in the Former Hand Area," *Nature Communications* 4 (2013): 1570; T. R. Makin et al., "Reassessing Cortical Reorganization in the Primary Sensorimotor Cortex Following Arm Amputation," *Brain* 138 (2015): 2140–2146.

16. Flor, Nikolajsen, and Staehelin Jensen, "Phantom Limb Pain: A Case of Maladaptive CNS Plasticity?"

17. Maguire et al., "Navigation-Related Structural Change in the Hippocampi of Taxi Drivers"; Maguire, Woollett, and Spiers, "London Taxi Drivers and Bus Drivers"; Woollett and Maguire, "Navigational Expertise May Compromise Anterograde Associative Memory"; Woollett, Spiers, and Maguire, "Talent in the Taxi."

18. Maguire, Woollett, and Spiers, "London Taxi Drivers and Bus Drivers"; Woollett and Maguire, "Navigational Expertise May Compromise Anterograde Associative Memory"; Woollett and Maguire, "Acquiring 'the Knowledge' of London's Layout Drives Structural Brain Changes."

19. Woollett, Spiers, and Maguire, "Talent in the Taxi."

Tool Use Can Instantly Rewire the Brain

Alison L. Barth

YOU'RE DRIVING HOME from your Saturday errands and you pull into the garage, turning at just the right moment to avoid the trash cans on one side and the bicycles on the other. You pull up just enough to make sure the back end of the car will clear the garage door as it closes. How are you so confident about where the edges of your car are? A fascinating aspect of driving is that when you sit at the steering wheel, your sense of "body" expands to become the outside of the car. You know how close you can drive to the curb and not rub up against it, you understand how far you can pull back without hitting an obstacle, and you know the relative position of your car in the road by glancing at the dotted yellow line. Your perception of "self" becomes bounded by the bumpers and steel of the car, increasing in volume more than fortyfold.

Remarkably, when you unbuckle your seatbelt and step outside of the car, your sense of body, also called somatosensation, instantly shrinks back to the boundaries of your skin. This rapid expansion and contraction of body sense reflect the striking plasticity of the brain. Plasticity is a fundamental property of the brain, and scientists are discovering the precise mechanisms by which specific neural pathways in somatosensory brain areas are weakened, strengthened, or masked to enable us to feel and do incredibly sophisticated things. Here, I will argue that the plasticity in the way that our body is represented in the brain occurs continuously for us as we move through our daily life. This is particularly relevant to us as a tool-using species. Every time we pick up an object and use it in an expert way, we are expanding our bodily representation to encompass and control that tool.

Infants must learn where their bodies begin and end, a process accomplished by touching and moving and watching their limbs in action. In this way, a map of the body surface starts as a hard-wired, rough draft that is set in place during embryonic development, but it is then refined by experience. The term "somatosensory map" refers to the principle that the organization of tactile information across the surface of the brain roughly matches what you would find in the skin: the representation of the thumb is adjacent to that of the fingers and is contiguous with the rest of the hand, then the elbow, etc. Although the exact amount of brain space devoted to a body area can vary—for example, the fingers and the lips occupy a disproportionately large region compared to the knees or shoulders—local relationships are generally preserved.

As children grow, they often pass through an ungainly stage where they struggle to keep up with the changes in their size and shape—legs are longer than expected and the children trip and fall, or they climb into a familiar lap for comfort and realize that they can barely fit any more. These somatosensory maps, located in the brain region called the neocortex, must adjust to reflect their new outline in the world—but fortunately, there are mechanisms that are specialized for this adjustment to occur.

An important property of somatosensory maps is that they can be altered by experience. Overstimulation of some parts, like the fingers of a violinist who practices for six hours a day, can cause the relative area that represents those fingers in the neocortex to increase.[1] As body size changes—for example, with weight gain or loss—somatosensory maps must also shift. Every pregnant woman has had the experience of trying to squeeze through a crowd of people only to realize that she is now much larger than she predicted! After that, she must readjust her expectations for where she can comfortably fit. This day-to-day, experience-dependent plasticity in somatosensory representations happens routinely, not just under rarified laboratory conditions.

Somatosensory loss from catastrophic events, such as the amputation of a limb—or the removal of a tooth, a more common experience—

can also lead to the redistricting of brain territories, where intact sensory areas can eventually "take over" the cortical space vacated by the absent input. When the tooth is removed, our instinct is to continuously touch and feel that absent area, obsessively moving across the space. But within a few days or weeks, that empty space becomes again part of us, unnoticeable. The plasticity of neural connections in the brain is responsible for this ability to readjust our expectations about the size and shape of our bodies. Scientists are developing an increasingly sophisticated appreciation for the way that neurons in the brain can anatomically rewire based upon experience. In many cases, experience-driven rewiring can be stable and long-lasting. In other cases, neural circuits are rewired functionally, without a corresponding anatomical change, by changing the strength of synaptic connections between neurons. This is an efficient strategy—since it takes advantage of connections that are already present—and it enables rapid switching of somatosensory representations depending on the task at hand. It is also incredibly advantageous to us as a species.

Just as impressive as long-lasting adjustments in how we feel ourselves in the world—how we appreciate our size and shape—are cases where our somatosensory representations can be instantly adjusted depending on the task at hand. Think, for example, about when we start the car to pull out of a garage and immediately know when the side mirror is going to hit the wall. Or when we pick up a knife and fork. The edges of our body are immediately transformed to incorporate this tool so that it becomes part of us, extending the sensory reach of our fingers. Holding the fork, we can feel the plate and the morsel of food we pick up. Our "edges" extend to the tines of the fork, inches past the tips of our fingers. In fact, this form of experience-dependent plasticity in the somatosensory system may enable our expert use of tools, so the brain readjusts our sense of bodily boundaries, allowing us to use these tools as an extension of ourselves.

Why does expert tool use require practice? Because we need time and repetition—with feedback so that we learn by trial and error—to enable this remapping. That can be costly, as new drivers (or their parents) are dismayed to find, since it might take a scrape or two to figure out how far exactly one must be from another car to avoid an accident. But in fact, our brains are exquisitely prepared to accommodate new somatosensory inputs, extending the size and shape of our bodily edges to enable us to acquire new skills. The feedback we receive from trying to use chopsticks is in getting them to do what we intend. When our intentions become reflected in successful actions, the remapping of our body space becomes strengthened and consolidated. Other studies indicate that such trial and error periods, punctuated by success and repeated trials separated by sleep, are particularly important in enabling neural plasticity.[2] Over time, we move from a clumsy attempt to use two sticks to the sensation that the chopsticks are an extension of our hands. This change in mental representation with expertise was first described by psychologists over one hundred years ago and has been well characterized by many others since.[3] This constant process of experience-dependent brain remapping has happened to us since we were toddlers learning how to wear our first pair of shoes, to school children learning how to use a pencil, in skiing or painting or flipping a burger or playing the piano. There are changes in the way that cells in the brain respond to inputs that underlie this expertise—some of which are restricted to parts of the brain that control movement but others that unquestionably occur in sensory areas.

The brain is composed of almost 100 billion neurons, and the properties of these neurons can be changed by experience, disease, injury, or drugs. There are so many neurons in the brain that it can be hard to decide which ones we need to examine. When we think about tool use, we can narrow our focus to neurons that are in a discrete area and are activated by tactile manipulation and correspond to specific body areas. We know a few things for certain: training with one hand does not easily transfer to another hand (or foot, for that matter);[4] expertise requires practice (suggesting long-term changes in neural wiring properties); and the sense of tool incorporation into our body schema can be nearly instantly reversed. This reversal is apparent when we step out of the car or put down the fork—the body's edges snap back to their original state, suggesting that the changes in wiring properties can be activated or masked in a situation-dependent manner. Our everyday experience tells us that we can master many different types of tools, and this implies that these "tool maps" must coexist and probably overlap.

Experimental work in animals shows that visual feedback can aid the expansion of neural response properties, where the tool becomes part of the body representation. One brain area implicated in this process in nonhuman primates is called the intraparietal sulcus, which can combine visual and somatosensory information.[5] We need to understand how different types of neurons—both excitatory and inhibitory varieties—and the connections between them are dynamically changed by expert tool use. Without this understanding, it will be hard to develop an explanation of how biological components of the brain can give rise to somatosensory plasticity, let alone harness it for recovery and repair of brain function.

It is very likely that the normal mechanisms of plasticity and sensory memories are coopted to enable expert tool use, in the same way that repeated tactile input can drive changes in neural firing in the neocortex.[6] This process almost certainly includes an increase in the strength of connections between excitatory neurons in somatosensory areas[7] as well as changes in motor brain areas.[8] However, long-lasting change in excitatory neural connections would not be enough to explain the rapidity by which we can pick up and use a tool, switch between tools, or return to our original, naked, tool-less state. After all, the critical aspect of this situation-dependent expansion of what we consider "self" is that it can be instantly reversed. Thus the brain must have an ability to mask these strengthened connections through inhibition—when we put down the spatula, get out of the car, or take off our boots.

It remains mysterious how maps altered by tool use can remain separated from each other so that we can use different and varied objects without confusing, for example, a hammer with a pair of tweezers. Are there methods that we could highjack to enhance the acquisition of expert use? How is the natural variation among individuals manifested in the skills—are some people better at acquiring certain skills and why? All these questions remain active areas of research. One thing is clear: our brains were built by ancient evolutionary processes that did not anticipate that we would pick up objects in our environment to extend our physical abilities. Whatever normal cellular and synaptic mechanisms for experience-dependent plasticity existed in the central nervous system can be adapted for new purposes to enable us as a species to achieve ever more complicated skills. As we enter into an era where virtual reality becomes commonplace, we may discover new ways to reorganize our perceptual capacities, not just limited to how we use an object to turn it into a tool. Computer-aided feedback might enable us to speed up acquisition of skilled tool use, and tactile feedback in virtual reality will transform the rather flat visual experience to which it is currently restricted. New physical laws (reduced gravity?), imaginary objects, or altered timescales in interactive computer games will expand the physical world for us, leading to brain remapping in ways that the real world cannot initiate.

NOTES

1. T. Elbert, C. Pantev, C. Wienbruch, B. Rockstroh, and E. Taub, "Increased Cortical Representation of the Fingers of the Left Hand in String Players," *Science* 270 (1995): 305–307.

2. S. Diekelmann and J. Born, "The Memory Function of Sleep," *Nature Reviews Neuroscience* 11 (2010): 114–126.

3. M. Martel, L. Cardinali, A. C. Roy, and A. Farne, "Tool-Use: An Open Window into Body Representation and Its Plasticity," *Cognitive Neuropsychology* 33 (2016): 82–101.

4. J. A. Harris, I. M. Harris, and M. E. Diamond, "The Topography of Tactile Learning in Humans," *Journal of Neuroscience* 21 (2001): 1056–1061.

5. A. Iriki, M. Tanaka, and Y. Iwamura, "Coding of Modified Body Schema during Tool Use by Macaque Postcentral Neurones," *Neuroreport* 7 (1996): 2325–2330; S. Obayashi et al., "Functional Brain Mapping of Monkey Tool Use," *Neuroimage* 14 (2001): 853–861.

6. M. E. Diamond, W. Huang, and F. F. Ebner, "Laminar Comparison of Somatosensory Cortical Plasticity," *Science* 265 (1994): 1885–1888; S. Glazewski and A. L. Barth, "Stimulus Intensity Determines Experience-Dependent Modifications in Neocortical Neuron Firing Rates," *European Journal of Neuroscience* 41 (2015): 410–419; S. Glazewski and K. Fox, "Time Course of Experience-Dependent Synaptic Potentiation and Depression in Barrel Cortex of Adolescent Rats," *Journal of Neurophysiology* 75 (1996): 1714–1729.

7. R. L. Clem and A. Barth, "Pathway-Specific Trafficking of Native AMPARs by in vivo Experience," *Neuron* 49 (2006): 663–670.

8. M. S. Rioult-Pedotti, D. Friedman, G. Hess, and J. P. Donoghue, "Strengthening of Horizontal Cortical Connections Following Skill Learning," *Nature Neuroscience* 1 (1998): 230–234; G. Yang, F. Pan, and W. B. Gan, "Stably Maintained Dendritic Spines Are Associated with Lifelong Memories," *Nature* 462 (2009): 920–924.

Life Experiences and Addictive Drugs Change Your Brain in Similar Ways

Julie Kauer

WHY CAN'T WE REMEMBER our most wonderful experiences as vividly as we would like? Why can't we quickly forget something painful no matter how hard we try? A memory has its own time course, fading slowly over time, whether we like it or not. Surprisingly, drug addiction has features similar to memory. If a substance abuser tries to quit, he or she faces a problem analogous to trying hard to forget a bad experience: a lack of voluntary control over the drug-associated memories that drive relapse. Why does addiction have this mnemonic character?

Every day you learn and experience new things. Some memories are quickly forgotten (where you parked your car yesterday morning), while others are remembered. Even if the new experience is so commonplace that you barely notice—let's say you see your neighbor driving a new car—that information is still stored for later retrieval. The only way this retrieval can happen is if your brain subtly changes as you store the memory of the new car. The newly rewired brain has incorporated the information so that later you can retrieve this new fact.

Memories are formed in the brain by a process of strengthening and weakening synapses, the connections between individual neurons. As a result of synaptic strengthening, synapses between two neurons will subsequently more strongly drive electrical activity in the receiving cell in the circuit. From personal experience alone, we can identify some features of learning and memory that appear to be encoded in the brain by synaptic plasticity. First, we can learn very rapidly. If we meet someone new, it takes only seconds to encode a memory of that moment and the person's face and name. Second, some memories are more fleeting than others. For example, after meeting someone we may remember the face, but the name may escape us the next time we meet. Third, life events that are particularly important or emotionally charged can be remembered for a long time in exquisite detail. The first day of school, the day we bought our first guitar, the day a child was born—these and other critical moments are laid down in memory immediately and persist for years. Salient memories like these can also be difficult or impossible to erase. The memory of what we were doing on 9/11 or the day a hurricane hit can stay with us for years, even if we want nothing better than to forget them.

The rewiring of synapses through changes in synaptic strength (synaptic plasticity) shares and can account for the properties we recognize in learning and memory formation. Synaptic rewiring takes place within seconds. Some synaptic changes last longer than others, and synaptic plasticity can be highly persistent, lasting long enough to account for long-lasting memories. These synaptic changes are localized in specific brain regions, such as the hippocampus, that are known to be required for learning and for encoding memory.

Remarkably, a nearly identical brain-rewiring process occurs if you take an addictive drug. Like memory, the development of drug addiction is also caused by brain changes.[1] Perhaps this is obvious, but it is worth thinking about. Even taking a drug a single time persistently changes the way your brain works, thereby altering the way you experience the world thereafter. And while every addictive drug changes synaptic strength, antidepressants like Prozac or drugs used to treat epilepsy target the brain but are not addictive. Unlike addictive drugs, antidepressants and antiepileptic drugs do not release dopamine or promote synaptic changes in the brain's motivational circuitry, and this may explain their nonaddictive nature.[2]

Drugs of abuse act on specific target molecules in the brain and alter brain function rapidly and for long time periods after exposure.[3] These brain changes are the reason addiction is such a difficult problem to treat and reverse. The ventral tegmental area and nucleus accumbens regions of the brain comprise a dopamine-using neural circuit that can be thought of as a motivation center; these areas are active during motivated behavior, and motivated responses are lost if they are damaged. Strong evidence for this is that damage to the nucleus accumbens, for example, but not to other brain regions, disrupts addiction to nicotine in human smokers.[4] A determination to stop being addicted is as much an uphill battle as being determined to forget a bad memory—not impossible but very difficult. Substance abusers experience drug craving, an inability to do or think about anything else. Craving is exacerbated by anything associated with the previous drug use; if you always smoked a cigarette after lunch, then once you quit, in the period after lunch you may find your mind obsessively preoccupied with the impulse to smoke. Craving and associated compulsive thoughts are not unique to drug addiction. Imagine you are out on a long hike without a water bottle. As the sun beats down, thoughts of water—or, perhaps more accurately, compelling and unpleasant feelings of needing water (craving)—intrude more and more often on your internal dialogue, interrupting thinking about anything else. No beautiful sunset or delicious slice of pizza can stop this unpleasant intrusive sensation until your thirst is quenched.

It's easy to see why the nervous system evolved a motivational circuit to create this uncomfortable state during extreme thirst. Without water, we die within a few days, so it is critical that there is a brain circuit responsible for sending urgent, intrusive reminders of our survival needs. Drugs of abuse alter the same brain circuits. The drive to seek out life-sustaining environmental cues like food and water is maintained because these substances activate the central players in the brain motivation circuit, dopamine-using neurons of the ventral tegmental area. These neurons manufacture the neurotransmitter dopamine and release it onto their downstream target cells in brain regions like the nucleus accumbens that are also important components of the motivation circuit. Dopamine neuron firing appears to signal the things we urgently need to survive, and dopamine cells become active in response to food, water, warmth, and even sex. What if a chemical, either found in nature or cooked up in a lab, could tap into the motivational circuit and drive dopamine neurons artificially from within the brain? Intriguingly, this may be exactly how drugs of abuse work. Although different drugs of abuse have distinct molecular targets and very different behavioral effects, they all drive the electrical activity of dopamine neurons or the release of dopamine from these cells (while nonaddictive brain-targeted drugs like Prozac do not).[5]

Although dopamine is released for only about a half hour during eating or drinking, once addictive drugs access the brain, they can raise dopamine levels for hours.[6] Initially, this persistent deluge of dopamine can feel very rewarding, at least as good as a cool drink after a long, hot hike. Perhaps more important, over the long term, when addictive drugs like cocaine or oxycontin reach the motivational circuit, the brain appears to interpret the rush of dopamine as the signal for a substance essential for survival. Even when taken for the first time, cocaine or oxycontin usurp the existing motivational circuit by strengthening synapses that drive dopamine cells and their downstream target cells.[7] In this situation, the result of drug-evoked synaptic plasticity is a big increase in motivation to seek and take the drug rather than encoding a new memory. If the motivational circuit is a finely tuned system optimized to identify pivotal environmental events by making them feel good and reinforcing the desire for them, then rewiring this circuit can have disastrous consequences. During addiction, everything in life becomes subordinated to the desire to obtain and use the drug, just as if the drug were critical for survival.

When lab animals are hungry or thirsty, they will press a lever to receive food or water. They will also press a lever to receive any of the drugs of abuse that humans will take, even if they must work extremely hard to receive them.[8] This observation emphasizes that in experimental work, rodents can be profitably studied to provide insights into how addictive drugs affect the brain in ways relevant to humans. Animal experiments have demonstrated when and how addictive drugs trigger changes in brain synapses in the motivational circuit. Synaptic strengthening takes place at several time points after drug exposures and at distinct synaptic locations within the motivational circuit. In the dopamine-using cells of the ventral tegmental area, plasticity occurs within a few hours of drug exposure, like a rapid new learning event. It is important that the same synapses are also strengthened after food or sugar ingestion in hungry rodents, but they only remain stronger for a few days. Instead, cocaine triggers rapid synaptic plasticity lasting for months.[9]

This rapid rewiring could be viewed as a form of learning, though it is unconscious, taking place without any sense of the brain's alteration. Though this rapid plasticity does not equal addiction—the great majority of drugs are not addictive until they have been used repeatedly—perhaps these early synaptic events in the ventral tegmental area indicate a lowering of the threshold for drug dependence, or they may mark the drug-taking experience as a highly salient event when next encountered.[10] In contrast, in the nucleus accumbens a single drug exposure does nothing to synapses, and plasticity instead develops only after repeated drug administration. And once a person is drug-experienced and tries to quit, an insidious form of plasticity sets in, known as incubation of craving.[11] While craving begins early during an abstinent period, with the passage of weeks off the drug the craving becomes more and more exaggerated. Paralleling this behavioral change, during abstinence synapses in the nucleus accumbens downstream from dopamine neurons become stronger and stronger.[12] Once this synaptic plasticity has occurred, it does not seem to diminish but instead intensifies and stabilizes over time, something like a long-term memory. Your brain after repeated use of drugs is truly different from your brain before drugs.

Although much remains unknown about brain changes during addiction, synaptic alteration of the motivational circuit that follows drug use is one likely culprit. What are the consequences for the substance abuser who has quit and stayed clean? Is there no way back to the normal brain? Synaptic plasticity in motivational circuits provides one explanation for why quitting and remaining drug free are so difficult to achieve. The brain after drugs has a modified motivational circuit, and the reversal of synaptic strengths to the previous predrug starting point is as difficult to achieve as a simple reversal and loss of a specific memory. However, just as a drug can change the brain rapidly and indelibly, the very plasticity of its synaptic wiring means that the brain is also amenable to therapeutic remodeling. There are already examples from animal studies of addictive drugs of how synapses can be weakened or motivational circuits remodeled by experimental intervention. For example, drug treatments have been identified in rodents that reverse synaptic strengthening processes resulting from cocaine exposure.[13] Electrical stimulation of the brain is also being used to treat neurological diseases, including Parkinson's disease,[14] and similarly driving specific locations in the motivational circuit using deep brain stimulation and other new technologies offer the promise of beneficial rewiring of the addicted brain even without medication.[15] Such work holds out hope that new drugs and other treatments will provide solid therapeutic avenues to reversing the devastating behavioral consequences of addiction.

NOTES

1. R. Z. Goldstein and N. D. Volkow, "Dysfunction of the Prefrontal Cortex in Addiction: Neuroimaging Findings and Clinical Implications," *Nature Reviews Neuroscience* 12 (2011): 652–669.

2. D. Saal, Y. Dong, A. Bonci, and R. C. Malenka, "Drugs of Abuse and Stress Trigger a Common Synaptic Adaptation in Dopamine Neurons," *Neuron* 37 (2003): 577–582.

3. C. Luscher, "Drug-Evoked Synaptic Plasticity Causing Addictive Behavior," *Journal of Neuroscience* 33 (2013): 17641–17646.

4. N. H. Naqvi and A. Bechara, "The Insula and Drug Addiction: An Interoceptive View of Pleasure, Urges, and Decision-Making," *Brain Structure and Function* 214 (2010): 435–450.

5. G. Di Chiara and A. Imperato, "Drugs Abused by Humans Preferentially Increase Synaptic Dopamine Concentrations in the Mesolimbic System of Freely Moving Rats," *Proceedings of the National Academy of Sciences* 85 (1988): 5274–5278.

6. G. Di Chiara and V. Bassareo, "Reward System and Addiction: What Dopamine Does and Doesn't Do," *Current Opinions in Pharmacology* 7 (2007): 69–76.

7. J. A. Kauer and R. C. Malenka, "Synaptic Plasticity and Addiction," *Nature Reviews Neuroscience* 8 (2007): 844–858.

8. V. Deroche-Gamonet, D. Belin, and P. V. Piazza, "Evidence for Addiction-like Behavior in the Rat," *Science* 305 (2004): 1014–1027.

9. B. T. Chen, M. S. Bowers, M. Martin, F. W. Hopf, A. M. Guillory, R. M. Carelli, J. K. Chou, and A. Bonci, "Cocaine but Not Natural Reward Self-Administration Nor Passive Cocaine Infusion Produces Persistent LTP in the VTA," *Neuron* 59 (2008): 288–297.

10. J. A. Kauer, "Learning Mechanisms in Addiction: Synaptic Plasticity in the Ventral Tegmental Area As a Result of Exposure to Drugs of Abuse," *Annual Review of Physiology* 66 (2004): 447–475.

11. L. Lu, J. W. Grimm, B. T. Hope, and Y. Shaham, "Incubation of Cocaine Craving after Withdrawal: A Review of Preclinical Data," *Neuropharmacology* 47 (2004): 214–226.

12. J. A. Loweth, K. Y. Tseng, and M. E. Wolf, "Adaptations in AMPA Receptor Transmission in the Nucleus Accumbens Contributing to Incubation of Cocaine Craving," *Neuropharmacology* 76 (2014): 287–300.

13. M. Mameli, B. Balland, R. Lujan, and C. Luscher, "Rapid Synthesis and Synaptic Insertion of GluR2 for mGluR-LTD in the Ventral Tegmental Area,"

Science 317 (2007): 530–533; J. E. McCutcheon, J. A. Loweth, K. A. Ford, M. Marinelli, M. E. Wolf, and K. Y. Tseng, "Group I mGluR Activation Reverses Cocaine-Induced Accumulation of Calcium-Permeable AMPA Receptors in Nucleus Accumbens Synapses via a Protein Kinase C-Dependent Mechanism," *Journal of Neuroscience* 31 (2011): 14536–14541.

14. A. V. Kravitz, D. Tomasi, K. H. LeBlanc, R. Baler, N. D. Volkow, A. Bonci, and S. Ferre, "Cortico-Striatal Circuits: Novel Therapeutic Targets for Substance Use Disorders," *Brain Research* 1628 (2015): 186–198.

15. Y. Y. Ma, X. Wang, Y. Huang, H. Marie, E. J. Nestler, O. M. Schluter, and Y. Dong, "Re-silencing of Silent Synapses Unmasks Anti-Relapse Effects of Environmental Enrichment," *Proceedings of the National Academy of Sciences of the USA* 113 (2016): 5089–5094.

SIGNALING

Like It or Not, the Brain Grades on a Curve

Indira M. Raman

HAPPINESS, in one form or another, seems to be a common goal that most of us would like to attain. We often behave as though we might find a route to contentment—comfort, satiety, warmth, or some other reward—and be happy all the time if we could just make the right choices. But pleasure is often fleeting, even from the most appealing experiences, giving rise to ennui and sparking the drive for something new and *sensational*. As a neuroscientist, I can't help wondering whether the transience of our satisfaction may not in fact be inescapable and instead may reveal an inevitable aspect of the way the brain works, the understanding of which might provide a clue to how to contend with it.

Many moment-to-moment functions of the brain seem so natural that we can hardly distance ourselves enough to reflect upon them: the brain *notices*. It is obvious, once we consider it, that a basic job of the brain is to perceive; with those perceptions it can evaluate; and based on those evaluations, it can act. This work is carried out by neurons of the nervous system. They detect and represent input from the outside world (and the inside world), analyze the data, and then respond to this analysis with an appropriate action. Action generally involves movement: neurons send the signals that make muscles contract and let you do things. The input is *sensory*, the analysis is often called *associative*, and the output is *motor.* The sensory-associative-motor triplet is the neural version of *perceiving, evaluating,* and *doing.*

How do the neurons that compose the brain conduct the business of detecting and analyzing what is happening out there in the world? The simplified answer is that they rely first on a translation service. The body parts that we think of as our sense organs—eyes, ears, noses, tongues, skin—contain sensory receptor cells, so called because they *receive* information. Tiny protein molecules sit on the membranes of those cells and translate (or, to use the technical term, transduce) physical stimuli from the outside world—light, sound, chemicals, and heat—into electrical signals called action potentials, which form the language of the brain. The transduction proteins form or connect to a minuscule pathway, or ion channel, through which charged particles called ions, like sodium and potassium, enter or exit the cell. The movement of ions makes the electrical signals. Each electrical signal spreads along the length of the cell by means of other proteins—which also form ion channels—ultimately culminating in the release of a chemical neurotransmitter. The next neuron receives the neurotransmitter via other receptor proteins, which also are themselves ion channels or are coupled to ion channels. Our ability to notice lies largely in our ion channel proteins.

The interesting part is that nearly all these proteins respond to *changes* in stimuli, but in the presence of long-lasting, constant stimulation of mild to moderate intensity, many of them quite literally shut themselves down and stop allowing ions to flow through them. We call this process adaptation (or desensitization or inactivation, depending on its physical basis). It leads to familiar experiences with respect to the senses. Adaptation is why, for example, when you go from a brightly lit space into a dim room, it seems dark at first, but after a while you no longer perceive darkness; the lighting seems normal. Only when you go back into the sun does the change make you realize how dim it was before—or how brilliant it is now. Similarly, most people adapt to the scent of cooking soon after entering a restaurant, or the coolness of a swimming pool after jumping into it on a hot day, or the background hum of a refrigerator. After a short exposure, the aroma or chill or noise—unless it is overwhelming to the point of discomfort—becomes undetectable and goes unnoticed. In the common parlance, you get used to it. In part because of our adapting ion channels, we perceive many things not by their absolute value but by their contrast to what came before.[1] In an extreme case, experimentalists have been able to demonstrate this phenomenon by stabilizing an image on the retina. Our eyes generally dart around in so-called microsaccades, which let our retinal cells compare the light reflected from dark and light areas in any visual scene. By monitoring a person's eye movements and shifting a projected image accordingly, visual neuroscientists can demonstrate that when an image is artificially held in a fixed position on the retina, the person "sees" the image disappear.[2] Without one's being able to make comparisons, the world goes gray. In other words, it's not just that variety is the spice of life; it's variance that lets us sense anything at all.

This sensitivity to change and insensitivity to constancy doesn't stop at the level of sensory receptors. Deeper in the brain, in almost every neuron, are other ion channel proteins—particularly sodium channels, which start action potentials (by letting sodium ions into a neuron), and potassium channels, which end action potentials (by letting potassium ions out of a neuron). Both sodium and potassium channels come in many varieties, and many of these ion channels also inactivate—that is, turn themselves off—with use. Consequently, even when chemical neurotransmitters provide a prolonged or repeated stimulus to neurons, the intrinsic properties of ion channels limit how many action potentials are produced. For instance, in some neurons, the inactivation of sodium channels makes it progressively more difficult for action potentials to be generated in the face of constant stimulation.[3] Meanwhile, specific potassium channels gradually *increase* their ion flow, helping to slow or shut down a neuron's signaling after a few action potentials. This interaction between sodium and potassium ion flow permits electrical signals to be generated only at the outset of a stimulus, a process called accommodation. Although there are exceptions, most of the principal excitatory cells of the cortex and hippocampus—the ones that encourage action potentials in the neurons they target—tend to accommodate.[4] We don't always know what kinds of information these accommodating neurons are carrying, but we do know that they respond most strongly to changing stimuli.[5]

Similarly, neurotransmitter receptor proteins can undergo desensitization, in which their ion channels shut themselves off in real time as prolonged stimuli arrive at the neuron.[6] But neurons also have an interesting ability to respond to long-term increases in neurotransmitter exposure—over days or longer timeframes which might result from excessive signaling through a particular neural circuit—by simply consuming their own neurotransmitter receptors, leaving fewer working receptors available on the cell surface. In part, such responses can underlie tolerance to medications, drugs of abuse, and even spicy food.[7] Conversely, when neurotransmitter release falls, a given neuron can produce more receptor proteins and associated ion channels. In this way, overstimulation reverts to producing a normal degree of input, and understimulation sets up a neural circuit to be extra sensitive even to small signals. How does a cell know? A variety of cellular feedback systems, many of which make use of the special biochemical properties of calcium ions, allows neurons to figure out, so to speak, the comfortable or appropriate point between too much and too little. Processes like these may be engaged when a stimulus that was initially pleasurable—or aversive—is experienced over and over again. The acute perception fades as the brain finds its own set point.[8]

At the level of the whole organism, the feelings from those perceptions fluctuate accordingly, diminishing to repeated stimuli and being restored only when a change takes place. A simple illustration of this phenomenon occurs in the sea slug *Aplysia,* which initially retracts its gill in response to a light touch. With a series of harmless touches, it habituates, ceasing to react, until the touch is paired with something more aversive, like a shock.[9] In a more pleasurable arena of sensation, hungry rats will expend effort to obtain either ordinary or palatable food, whereas rats that have eaten to satiety will work only to get new treats they find especially tasty. The rats' motivation to work for edibles can be reduced by drugs that interfere with receptors for natural opiates and dopamine, which are neurotransmitters in brain circuits that signal rewards. The idea, therefore, is that reward pathways are stimulated by the anticipation and/or consumption of food, but for sated rats only if the food compares favorably with their recent experience.[10] In other words, there is no need to save room for dessert; it will be just as pleasurable as long as it is better than what came before.

Familiar stimuli and the experiences they generate can also trigger other modifications of ion channels and neurotransmitter receptors, and these modifications can alter whole neural circuits. In fact, certain circuits in the brains of many animals (including us) are so good at predicting the outcome of well-known stimuli that they send inverse signals that actively cancel out the perception of what is going on. The organism doesn't even notice what is happening—at least until something different or surprising intervenes.[11] The ability to get used to and ultimately ignore incoming information that is static, familiar, predictable, and nonharmful turns out to be helpful behaviorally; in other words, it offers an evolutionary advantage. Continuing to notice sensations like the light touch of our clothes on our arms or the mild fragrance of the laundry detergent we used to wash them would be distracting, to say the least, and might even interfere with our ability to detect and respond to a signal that mattered, like a tap on the shoulder or our toast burning. In fact, an inability to predict and thereby adapt may be a contributing factor to conditions like autism spectrum disorders.[12] Besides, it's wasteful to send brain signals to report information we already know about. When all those ions flow in and out of cells to send signals within our brains, they cannot just remain on the opposite side from where they started. It literally consumes energy to pump sodium back out of neurons and potassium back into them, so it is most efficient not to generate action potentials that don't carry worthwhile information.

Does this mean that only novelty matters and that everything familiar must be discarded once the experience wears off? On the contrary; I think it offers a key to happiness that is compatible with how the brain works. The ability to detect even familiar stimuli can usually be restored by a brief palate cleanser, which literally permits a recovery from desensitization sufficient to intensify a subsequent experience. It is hard for me to assess how much I am waxing poetic, but it seems to me that the brain's ability—its need—to perceive by contrast may partly explain why our efforts to achieve perennial satisfaction have been largely unsatisfactory. Because the brain grades on a curve, endlessly comparing the present with what came just before, the secret to happiness may be unhappiness. Not unmitigated unhappiness, of course, but the transient chill that lets us feel warmth, the sensation of hunger that makes satiety so welcome, the period of near-despair that catapults us into the astonishing experience of triumph. The route to contentment is through contrast.

NOTES

1. Not all behavioral adaptation comes from ion channels that change rapidly, on the timescale of the stimulus. Many other long-term changes can take

place, some of which may involve ion channel function or expression (existence) and some of which tap into a variety of cellular processes. Also, not
every sensation or perception is subject to adaptation, one of the most
familiar sensations being the experience of pain.

2. R. W. Ditchburn and B. L. Ginsborg, "Vision with a Stabilized Retinal Image,"
   Nature 170 (1952): 36–37; S. Martinez-Conde, S. L. Macknik, X. G. Troncoso,
   and T. A. Dyar, "Microsaccades Counteract Visual Fading during Fixation,"
   Neuron 49 (2006): 297–305.

3. In contrast, some neurons have specialized sodium channels in which
   inactivation is circumvented by the intervention of an additional protein that
   literally blocks the usual inactivation process. These neurons readily fire long
   trains of high-frequency action potentials. Many such neurons are found in
   the cerebellum and brainstem; A. H. Lewis and I. M. Raman, "Resurgent
   Current of Voltage-Gated Na$^+$ Channels," *Journal of Physiology* 592 (2014):
   4825–4838.

4. In some cells, accommodation can be reversed by neurotransmitters like
   noradrenaline (norepinephrine), which suppresses the ionic current through
   specific potassium channels, called SK channels. It is interesting that the
   global effect on the brain of noradrenaline is often an increase in attention.
   Many toxins and poisons, such as those of scorpions and snakes, however,
   also prevent inactivation of sodium channels and block potassium channels,
   leading to convulsions and death, again revealing that the brain can suffer
   from too much of a good thing; D. V. Madison and R. A. Nicoll, "Actions of
   Noradrenaline Recorded Intracellularly in Rat Hippocampal CA1 Pyramidal
   Neurones, in vitro," *Journal of Physiology* 372 (1986): 221–244; B. Hille, "A K$^+$
   Channel Worthy of Attention," *Science* 273 (1996): 1677.

5. Although it is hard to resist the tempting conclusion that more brain activity
   is necessarily better, the ability of some neurons to shut down their own
   neuronal signaling by ion channel inactivation is often a good idea. A variety
   of neurological diseases is associated with too many action potentials in
   neurons that normally signal relatively sparsely. These disorders of "hyperex-
   citability" include some pain syndromes as well as epilepsy. The former yields
   too much sensation, and the latter yields too much muscle contraction; the
   symptoms depend on which classes of neurons get hyperactive. Often, the
   best medications for such conditions promote the inactivation of sodium
   channels. Even people without pain syndromes may be familiar with the
   pain-relieving effect of blocking sodium channels by their experience with
   Novocain at the dentist's office or lidocaine cream on sunburn. Medications
   for epilepsy are tailored not to shut down neural activity altogether but to
   constrain hyperactive neurons to accommodate.

6. Neurotransmitter receptors can shut down rapidly by desensitization, which
   is intrinsic to the protein, or by a short lifetime of the neurotransmitter itself,
   as it is destroyed by enzymes or soaked up by neighboring glial cells. Drugs
   and toxins that interfere with these processes and prolong the action of

neurotransmitters can have dramatic effects on the nervous system. Benzodi-azepines and other anxiolytic drugs extend the duration of ion flow through channels opened by the inhibitory neurotransmitter GABA. Nerve gas pro-longs the action of acetylcholine, the neurotransmitter that makes muscles contract.

7. The detection of spicy food is not done by neurotransmitter receptors in the brain but by chemical receptors at the periphery that respond to capsaicin, the naturally occurring chemical that makes chili peppers hot and painful. In an interesting twist on drug tolerance, capsaicin can be used as an ointment to desensitize and internalize receptors and relieve pain associated with condi-tions like arthritis and neuropathy.

8. This process is referred to as homeostasis, and much work has been directed toward studying "homeostatic plasticity" in neural circuits—the process by which neurons restore a basic set point of activity, even as the strength of the inputs that stimulate them varies; G. Turrigiano, "Homeostatic Synaptic Plasticity: Local and Global Mechanisms for Stabilizing Neuronal Function," *Cold Spring Harbor Perspectives in Biology* 4 (2012): a005736.

9. In the case of habituation, neurotransmitter receptors don't desensitize, but the neurons releasing chemical neurotransmitters run out of neurotrans-mitter; E. R. Kandel and J. H. Schwartz, "Molecular Biology of Learning: Modulation of Transmitter Release," *Science* 218 (1982): 433–443.

10. M. F. Barbano and M. Cador, "Opioids for Hedonic Experience and Dopa-mine to Get Ready for It," *Psychopharmacology* (Berlin) 191 (2007): 497–506.

11. An interesting illustration of a brain's ability to disregard the familiar comes from electric fish, which have an electric sense that lets them detect electric fields. These fish actively explore their environment by emitting a signal called an electric organ discharge (EOD)—the fish's own stereotyped "call," which produces an electric field around the fish. If an object is in the vicinity, the electric field is distorted—perhaps loosely analogous to feeling the distor-tion of your skin when you press on an object. It's the deviation of the signal from the usual that indicates the possibility of something worth fleeing from or investigating. The fish's own EOD signals nothing of potential signifi-cance. Accordingly, neurons that generate the EOD also send signals within the fish brain indicating that they have done so. The signal is exactly the opposite of the sensory input that the fish receives from its own undistorted EOD, effectively neutralizing the fish's sensation of its own "call" when there is nothing to be detected; C. Bell, D. Bodznick, J. Montgomery, and J. Bastian, "The Generation and Subtraction of Sensory Expectations within Cerebellum-like Structures," *Brain Behavior and Evolution* 50 (1997): 17–31.

12. M. Gomot and B. Wicker, "A Challenging, Unpredictable World for People with Autism Spectrum Disorder," *International Journal of Psychophysiology* 83 (2012): 240–247.

The Brain Achieves Its Computational Power through a Massively Parallel Architecture

Liqun Luo

THE BRAIN IS COMPLEX; in humans it consists of about 100 billion neurons, making on the order of 100 trillion connections. It is often compared with another complex system that has enormous problem-solving power: the digital computer. Both the brain and the computer contain a large number of elementary units—neurons and transistors respectively—that are wired into complex circuits to process information conveyed by electrical signals. At a global level, the architectures of the brain and the computer resemble each other, consisting of largely separate circuits for input, output, central processing, and memory.[1]

Which has more problem-solving power—the brain or the computer? Given the rapid advances in computer technology in the past decades, you might think that the computer has the edge. Indeed, computers have been built and programmed to defeat human masters in complex games, such as chess in the 1990s and recently Go, as well as encyclopedic knowledge contests, such as the TV show *Jeopardy!* As of this writing, however, humans triumph over computers in numerous real-world tasks—ranging from identifying a bicycle or a particular pedestrian on a crowded city street to reaching for a cup of tea and moving it smoothly to one's lips—let alone conceptualization and creativity.

So why is the computer good at certain tasks whereas the brain is better at others? Comparing the computer and the brain has been instruc-

Table 1

Comparing the computer and the brain

PROPERTIES	COMPUTER[a]	HUMAN BRAIN
Number of Basic Units	up to 10 billion transistors[b]	~100 billion neurons; ~100 trillion synapses
Speed of Basic Operation	10 billion/sec.	< 1,000/sec.
Precision	1 in ~4.2 billion (for a 32-bit processor)	~1 in 100
Power Consumption	~100 watts	~10 watts
Information Processing Mode	mostly serial	serial and massively parallel
Input/Output for Each Unit	1–3	~1,000
Signaling Mode	digital	digital and analog

[a] Based on personal computers in 2008.

[b] The number of transistors per integrative circuit has doubled every 18–24 months in the past few decades; in recent years the performance gains from this transistor growth have slowed, limited by energy consumption and heat dissipation.

References: John von Neumann, *The Computer and the Brain* (New Haven: Yale University Press, 2012); D. A. Patterson and J. L. Hennessy, *Computer Organization and Design* (Amsterdam: Elsevier, 2012).

tive to both computer engineers and neuroscientists. This comparison started at the dawn of the modern computer era, in a small but profound book entitled *The Computer and the Brain,* by John von Neumann, a polymath who in the 1940s pioneered the design of a computer architecture that is still the basis of most modern computers today.[2] Let's look at some of these comparisons in numbers (table 1).

The computer has huge advantages over the brain in the speed of basic operations.[3] Personal computers nowadays can perform elementary arithmetic operations, such as addition, at a speed of 10 billion operations per second. We can estimate the speed of elementary operations in the brain by the elementary processes through which neurons transmit information and communicate with each other. For example, neurons "fire" *action potentials*—spikes of electrical signals initiated near the neuronal cell bodies and transmitted down their long extensions called *axons*, which link with their downstream partner neurons. Information is encoded in the frequency and timing of these spikes. The highest frequency of neuronal firing is about one thousand spikes per second. As another example, neurons transmit information to their partner neurons mostly by releasing chemical *neurotransmitters* at specialized structures at axon terminals called *synapses*, and their partner neurons convert the binding of neurotransmitters back to electrical signals in a process called *synaptic transmission*. The fastest synaptic transmission takes about 1 millisecond. Thus both in terms of spikes and synaptic transmission, the brain can perform at most about a thousand basic operations per second, or 10 million times slower than the computer.[4]

The computer also has huge advantages over the brain in the precision of basic operations. The computer can represent quantities (numbers) with any desired precision according to the bits (binary digits, or 0s and 1s) assigned to each number. For instance, a 32-bit number has a precision of 1 in 2^{32} or 4.2 billion. Empirical evidence suggests that most quantities in the nervous system (for instance, the firing frequency of neurons, which is often used to represent the intensity of stimuli) have variability of a few percent due to biological noise, or a precision of 1 in 100 at best, which is millionsfold worse than a computer.[5]

The calculations performed by the brain, however, are neither slow nor imprecise. For example, a professional tennis player can follow the trajectory of a tennis ball after it is served at a speed as high as 160 miles per hour, move to the optimal spot on the court, position his or her arm, and swing the racket to return the ball in the opponent's court, all within a few hundred milliseconds. Moreover, the brain can accomplish all these tasks (with the help of the body it controls) with power consumption about tenfold less than a personal computer. How does the brain achieve that?

An important difference between the computer and the brain is the mode by which information is processed within each system. Computer tasks are performed largely in *serial* steps. This can be seen by the way engineers program computers by creating a sequential flow of instructions. For this sequential cascade of operations, high precision is necessary at each step, as errors accumulate and amplify in successive steps. The brain also uses serial steps for information processing. In the tennis return example above, information flows from the eye to the brain and then to the spinal cord to control muscle contraction in the legs, trunk, arms, and wrist.

But the brain also employs *massively parallel processing*, taking advantage of the large number of neurons and large number of connections each neuron makes. For instance, the moving tennis ball activates many cells in the retina called photoreceptors, whose job is to convert light into electrical signals. These signals are then transmitted to many different kinds of neurons in the retina in parallel. By the time signals originating in the photoreceptor cells have passed through 2–3 synaptic connections in the retina, information regarding the location, direction, and speed of the ball has been extracted by parallel neuronal circuits and is transmitted in parallel to the brain. Likewise, the motor cortex (part of the cerebral cortex that is responsible for volitional motor control) sends commands in parallel to control muscle contraction in the legs, the trunk, the arms, and the wrist, such that the body and the arms are simultaneously well positioned to receiving the incoming ball.

This massively parallel strategy is possible because each neuron collects inputs from and sends output to many other neurons—on the order of one thousand on average for both input and output for a mammalian neuron. (By contrast, each transistor has only three nodes for input and output all together.) Information from a single neuron can be delivered to many parallel downstream pathways. At the same time, many neurons that process the same information can pool their inputs to the same downstream neuron. This latter property is particularly useful for enhancing the precision of information processing. For example, information represented by an individual neuron may be noisy (say, with a precision of 1 in 100). By taking the average of input from 100 neurons carrying the same information, the common downstream partner neuron can represent the information with much higher precision (about 1 in 1,000 in this case).[6]

The computer and the brain also have similarities and differences in the signaling mode of their elementary units. The transistor employs *digital* signaling, which uses discrete values (0s and 1s) to represent information. The spike in neuronal axons is also a digital signal since the neuron either fires or does not fire a spike at any given time, and when it fires, all spikes are approximately the same size and shape; this property contributes to reliable long-distance spike propagation. However, neurons also utilize *analog* signaling, which uses continuous values to represent information. Some neurons (like most neurons in our retina) are nonspiking, and their output is transmitted by graded electrical signals (which, unlike spikes, can vary continuously in size) that can transmit more information than can spikes. The receiving end of neurons (reception typically occurs in the *dendrites*) also uses analog signaling to integrate up to thousands of inputs, enabling the dendrites to perform complex computations.[7]

- Another salient property of the brain, which is clearly at play in the return of service example from tennis, is that the connection strengths between neurons can be modified in response to activity and experience—a process that is widely believed by neuroscientists to be the basis for learning and memory. Repetitive training enables the neuronal circuits to become better configured for the tasks being performed, resulting in greatly improved speed and precision.

Over the past decades, engineers have taken inspiration from the brain to improve computer design. The principles of parallel processing and use-dependent modification of connection strength have both been incorporated into modern computers. For example, increased parallelism, such as the use of multiple processors (cores) in a single computer, is a current trend in computer design. As another example, "deep learning" in the discipline of machine learning and artificial intelligence, which has enjoyed great success in recent years and accounts for rapid advances in object and speech recognition in computers and mobile devices, was inspired by findings of the mammalian visual system.[8] As in the mammalian visual system, deep learning employs multiple layers to represent increasingly abstract features (e.g., of visual object or speech), and the weights of connections between different layers are adjusted through learning rather than designed by engineers. These recent advances have expanded the repertoire of tasks the computer is capable of performing. Still, the brain has superior flexibility, generalizability, and learning capability than the state-of-the-art computer. As neuroscientists uncover more secrets about the brain (increasingly aided by the use of computers), engineers can take more inspiration from the working of the brain to further improve the architecture and performance of computers. Whichever emerges as the winner for particular tasks, these interdisciplinary cross-fertilizations will undoubtedly advance both neuroscience and computer engineering.

NOTES

The author wishes to thank Ethan Richman and Jing Xiong for critiques and David Linden for expert editing.

1. This essay was adapted from a section in the introductory chapter of L. Luo, *Principles of Neurobiology* (New York: Garland Science, 2015), with permission.
2. J. von Neumann, *The Computer and the Brain* (New Haven: Yale University Press, 2012), 3rd ed.
3. D. A. Patterson and J. L. Hennessy, *Computer Organization and Design* (Amsterdam: Elsevier, 2012), 4th ed.
4. The assumption here is that arithmetic operations must convert inputs into outputs, so the speed is limited by basic operations of neuronal communication such as action potentials and synaptic transmission. There are exceptions to these limitations. For example, nonspiking neurons with electrical synapses (connections between neurons without the use of chemical neurotransmitters) can in principle transmit information faster than the approximately one millisecond limit; so can events occurring locally in dendrites.
5. Noise can reflect the fact that many neurobiological processes, such as neurotransmitter release, are probabilistic. For example, the same neuron may not produce identical spike patterns in response to identical stimuli in repeated trials.
6. Suppose that the standard deviation of mean (σ_{mean}) for each input approximates noise (it reflects how wide the distribution is, in the same unit as the mean). For the average of n independent inputs, the expected standard deviation of means is $\sigma_{mean} = \sigma / \sqrt{n}$. In our example, $\sigma = 0.01$, and $n = 100$; thus $\sigma_{mean} = 0.001$.
7. For example, dendrites can act as coincidence detectors to sum near synchronous excitatory input from many different upstream neurons. They can also subtract inhibitory input from excitatory input. The presence of voltage-gated ion channels in certain dendrites enables them to exhibit "nonlinear" properties, such as amplification of electrical signals beyond simple addition.
8. Y. LeCun, Y. Bengio, and G. Hinton, "Deep Learning," *Nature* 521 (2015): 436–444.

The Brain Harbors Many Neurotransmitters

Solomon H. Snyder

WHEN I JOINED the Johns Hopkins faculty in 1966, I met with the great neurophysiologist Vernon Mountcastle and told him about my previous work at the National Institutes of Health (NIH) studying different neurotransmitters and considering whether there existed substantial numbers of hitherto unidentified ones. He challenged me, "Why should the brain need more than a single excitatory and a single inhibitory transmitter?" I had no strong retort other than the sentiment that neurotransmission was substantially more nuanced than simple excitation or inhibition.

Before speculating on the number of neurotransmitters in the brain, we need an acceptable definition, at least a working version, of just what constitutes a neurotransmitter. Neuroscientists have been debating this issue for about a hundred years with no clear conclusions. The general consensus is that a candidate neurotransmitter should be a substance stored in neurons, released when they fire action potentials, and acting upon adjacent neurons or other cells (muscles, glia) to cause excitation or inhibition. Scientists vary in the rigor with which they apply these various criteria. For the purposes of this essay, substances stored in neurons, presumably released and exerting some sort of action on other cells, would qualify for the designation of "putative neurotransmitter."

I first encountered neurotransmitters during my training as a research associate at the NIH in Bethesda, Maryland, working with Julius Axelrod. Julie had recently been enmeshed in characterizing how some substances act as neurotransmitters, research for which he shared the 1970 Nobel Prize with Ulf von Euler and Bernard Katz. Von Euler had established that the molecule norepinephrine is the neurotransmitter of sympathetic neurons, the nerve cells that subconsciously coordinate the body's fight-or-flight response in potentially dangerous situations.[1] Bernard Katz had employed electrical recordings from nerves and muscles to show that a molecule called acetylcholine is stored in small ball-like structures, called synaptic vesicles, from which it is released in discrete packets, called quanta, each corresponding to the contents of a single vesicle.[2] Evidence for norepinephrine and acetylcholine as transmitters was rather solid, and another molecule called serotonin behaved so much like norepinephrine that most people would accept it as a neurotransmitter, even in the absence of definitive evidence.

Many neuroscientists were satisfied that these few substances, whose structures incorporated a nitrogen-containing entity called an amine, were the principal neurotransmitters. Others were dubious because most calculations suggested that the various amines could account for only a small percentage of synapses. An amino acid called GABA (gamma-aminobutyric acid) had been isolated by Eugene Roberts in the mid-1950s as a nervous system–specific substance that reduced neuronal firing and thus might be a major inhibitory neurotransmitter.[3] If you were to grind up a whole brain, you'd find that the concentration of GABA is roughly a thousand times greater than that of the neurotransmitter amines, implying that GABA might be one of the main neurotransmitters, at least in a quantitative sense. Visualizing the GABA synthesizing enzyme, called glutamic acid decarboxylase, in specific nerve endings convinced most investigators that GABA nerve terminals form a major portion of the total brain complement of synapses. To this day it is still difficult to judge what percentages of synapses in the brain are accounted for by one or another neurotransmitter. The best guesses for GABA have been in the range of 30–40 percent.[4] By contrast, the amines each comprise only a few percent. Numbers bandied about are roughly one percent each for norepinephrine, dopamine, and serotonin and perhaps 5 percent for acetylcholine.

The notion that GABA accounts for a major proportion of brain synapses and is inhibitory brought closer to closure the thinking about what proportions of brain synapses are represented by which types of transmitters. It was reasonable to suppose that there should be comparable numbers of excitatory and inhibitory synapses. Hence, if GABA synapses comprised 40 percent of the total, a comparable percentage should be accounted for by an excitatory transmitter or transmitters. In the 1950s and early 1960s a number of investigators had noted that the amino acids glutamate and aspartate are excitatory (they evoked spike firing) when applied to neuronal cells.

Just because glutamate can excite neurons does not mean that it does so in "real life." Concentrations of glutamate are among the highest of any substance in the brain. Neuroscientists were so entrenched in their belief that neurotransmitters should be "trace elements" that many refused to accept glutamate as a neurotransmitter simply because it was too abundant. Ideally one would like to visualize a candidate transmitter, such as glutamate, and ascertain whether it is stored in vesicles, and then count the number of terminals containing glutamate—a tall order. Nonetheless, there is now general agreement that amino acids, particularly glutamate and GABA, are the dominant brain transmitters, with glutamate the principal excitatory transmitter.

The rather small number of amines and amino acids that could be putative neurotransmitters was such that few scientists thought that the brain could harbor more than five or six transmitters. All of this changed with the report by Susan Leeman in 1970 that a new class of molecule called a peptide, substance P, occurs in the brain at high concentrations with properties suggesting a transmitter role.[5] In Sweden, a group of peptide chemists and pharmacologists began identifying dozens of peptides in the brain.[6] Some of these were already well known as hormones, acting largely in the gastrointestinal pathway, such as vasoactive intestinal polypeptide (VIP), cholecystokinin, gastrin, secretin, and somatostatin. Classic hormonal peptides such as insulin and glucagon became targets of speculation as potential transmitters. Pituitary peptides such as adrenocorticotropic hormone, growth hormone, and thyrotropin were shown to exist in selective neurons in the brain. The small class of hormones that act on the brain region called the hypothalamus, such as thryotropin-releasing hormone, corticotropin-releasing hormone, and growth hormone–releasing hormone, were also identified in neuronal populations.

A key figure in expanding the pool of peptide transmitters was the Swedish histochemist Tomas Hökfelt, who localized a wide range of peptides to unique neuronal groups. Peptides that had been well characterized outside the brain, such as angiotensin II, bradykinin, calcitonin, calcitonin gene-related peptide, neuropeptide Y, and galanin, have also been identified in neurons of the brain and are thought to be neurotransmitters.

A rigorous definition of a transmitter would include that it must be stored in synaptic vesicles, be released by fusion of those vesicles with a neuron's outer membrane, and act upon receptors on adjacent cells. However, the advent of the gaseous substance nitric oxide as a gasotransmitter overturned this rigid cluster of criteria.[7] There are no pools of nitric oxide in vesicles. As a labile gas, nitric oxide is synthesized by its biosynthetic enzyme, called nitric oxide synthase, from the amino acid arginine, "on demand." Background levels of nitric oxide are thought to be vanishingly low. The same considerations apply to the other gasotransmitters, carbon monoxide and hydrogen sulfide.[8]

Although the gasotransmitters are recent recruits to the family of neurotransmission, they may be of substantial importance both functionally and quantitatively. For instance, hydrogen sulfide signals by modifying the sulfhydryl (-SH) group on cysteine in protein targets, a process called sulfhydration.[9] About 35 percent of glyceraldehyde-3-phosphate dehydrogenase (GAPDH) molecules in the body are sulfhydrated, a condition that exerts a major influence upon the enzyme's activity in carbohydrate metabolism. Sulfhydration increases the activity of GAPDH severalfold. Thus its regulation impacts overall cellular disposition in a major way. In this way hydrogen sulfide may influence most cells in the body for diverse metabolic processes, not just neurotransmission.

This recounting provides hints at approaches to enumerating transmitters but nothing definitive. We are left with speculation. The great microbiologist Joshua Lederberg loved to conceptualize the "limits" of diverse biologic phenomena. I recall well a visit in his office in 1980, at a time that he was immersed in his hobby of astronomy and was trying to estimate the number of stars in the universe. According to his recounting, astronomers base their estimates on the number of known stars and the limitations of existing technology for identifying new ones. He challenged me to apply similar calculations to neurotransmitters, especially neuropeptides. In other words, we know of roughly fifty neurotransmitters. How many might exist that have escaped present identification mechanisms and/or the curiosity of today's scientists? I still do not have a reasonable answer. Moreover, why we need a large number of transmitters is an open question. Most neuroscientists feel that diverse transmitters each act in somewhat unique fashions. Some might simply excite or inhibit, at various timescales from milliseconds to tens of seconds. Others may trigger unique biochemical reactions that are not associated with electrical signals. Among the latter group there may be myriad different molecular pathways that mediate subtle neuronal alterations. A given transmitter might alter responses to other transmitters or act in a combinatorial fashion. One can speculate a broad range of distinct actions that would justify the existence of tens or even hundreds of transmitters.

In summary, how many putative neurotransmitters do I feel exist in the nervous system? Neurotransmitters occur in chemical classes: the amines, amino acids, peptides, and gases. As best as anyone can guess, numbers of amines, amino acids, and gases are notably constrained. By contrast, there could exist many more peptides than are presently known. Identifying most of these should not be an insuperable challenge. With presently available technology, one should be able to define the universe of biologically active peptides in all tissues and, especially, in the brain. I suspect the number will be finite, perhaps no more than two hundred. Hopefully, someone is already addressing this challenge, and it could provide a giant step forward in enhancing our appreciation of neurotransmission and, thereby, neural function.

NOTES

1. T. N. Raju, "The Nobel Chronicles, 1970: Bernard Katz (b 1911), Ulf Svante von Euler (1905–1983), and Julius Axelrod (b 1912)," *Lancet* 354 (1999): 873.
2. Ibid.
3. E. Roberts, "Gamma-Aminobutyric Acid and Nervous System Function—A Perspective," *Biochemical Pharmacology* 23 (1974): 2637–2649.
4. Ibid.
5. E. A. Mroz and S. E. Leeman, "Substance P," *Vitamins and Hormones* 35 (1977): 209–281.
6. T. Hökfelt, J. M. Lundberg, M. Schultzberg, O. Johansson, L. Skirboll, A. Anggård, B. Fredholm, B. Hamberger, B. Pernow, J. Rehfeld, and M. Goldstein, "Cellular Localization of Peptides in Neural Structures," *Proceedings of the Royal Society of London, Series B* 210 (1980): 63–77.

7. A. K. Mustafa, M. M. Gadalla, and S. H. Snyder, "Signaling by Gasotransmitters," *Science Signaling* 2 (2009): re2.

8. Carbon monoxide is formed from heme, best known as the oxygen-carrying
portion of hemoglobin, by a neuronally selective form of the enzyme called
heme oxygenase. This enzyme removes a one-carbon fragment from heme as
carbon monoxide. Hydrogen sulfide is generated from the amino acid cysteine
by one of two enzymes that were first described as acting upon a poorly understood amino acid derivative called cystathionine.

9. Mustafa, Gadalla, and Snyder, "Signaling by Gasotransmitters."

ANTICIPATING, SENSING, MOVING

The Eye Knows What Is Good for Us

Aniruddha Das

WE TEND TO THINK of vision as a super-video system that records everything with high fidelity. But in many ways the fidelity of vision is actually rather low: we are poor at judging absolute levels of brightness or spectral color. Yet we effortlessly gauge how fast a car is approaching (and, hopefully, jump out of the way) or reach for the one ripe mango in a pile and cup our hand properly in anticipation of grasping it. The key insight to understanding how vision works is that it is a purposeful biological process that has evolved to pick out information that would be most helpful for our survival and to ignore information that is not. And that it does so in a way that is well adapted to the particular properties of our visual world.

This evolved, purpose-driven visual processing starts in the eye.[1] To appreciate this, it is important to delve into the eye's anatomy and physiology just a bit. Light is focused onto the retina at the back of the eye, where it is converted into electrical responses in a set of photoreceptor cells—namely, rods and red-, green-, and blue-sensitive cones. This is the only aspect of the eye's function that can be usefully compared with that of a camera. Photoreceptors tile the retina with just the right spacing to best resolve the image formed by the eye's optical elements—notably the cornea and lens. And they each respond to the tiny patch of the image falling on them, much like pixels in a digital camera. But these individ-

97

ual pixel-like responses are not what is sent from the eye to the brain. Rather, they are first processed extensively within the retina by a complex multilayer network of neural circuits. It is this processed output that then gets sent out of the eye by the final elements of the retinal circuit, called retinal ganglion cells, whose output fibers, or axons, form the optic nerve carrying this electrical information to the next stage in the brain.

This image processing in the neural circuits of the retina is best described through the transformations of corresponding neural "receptive fields."[2] In the visual system, this term refers to the particular patch of visual space and pattern of visual stimulation that best drives a neuron. Consider recording the response—the change in cell membrane voltage—of a cone photoreceptor. The cell will respond only to light from the particular patch that this cone "sees" in visual space. The response is monotonic. Increasing the brightness of the light in the receptive field increases the cone response until it saturates. This tiny patch of visual space with its monotonic evoked response—and equivalent patches for each of the other roughly 120 million photoreceptors in the retina—define the receptive fields at the input level of the retinal circuit. At the output end, retinal ganglion cells combine the electrical responses of a few up to a few hundred individual photoreceptors. Light falling on the receptive fields of any of the cones providing input to the retinal ganglion cell will thus affect the cell's response voltage. But the effect is not monotonic and does not simply integrate over the input cones. Rather, the inputs are arranged in a circularly symmetric, antagonistic manner. Those in a central patch add up so that light falling on that patch changes the retinal ganglion cell voltage in one direction. Input from a ring-shaped surround is switched in sign and reverses the retinal ganglion cell voltage. The connections are also selective for cone type. Thus the retinal ganglion cell may respond positively to red in the center while being suppressed by green in the surround, or it may respond to blue in the center while being suppressed by yellow in the surround. This complex antagonistic center-surround response pattern defines the receptive field of the output retinal ganglion cell. One important class of such cells, called the "on-center magnocellular" cells, responds vigorously to a bright patch in the center and is suppressed by bright patches in the surround, independent of color. With a uniform wash or even a steady gradient of brightness over the full receptive field, the center and surround essentially cancel each other out, and

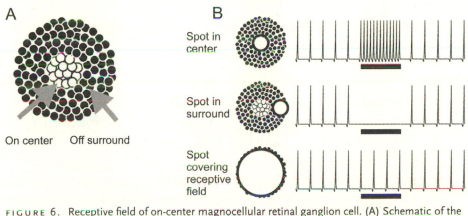

FIGURE 6. Receptive field of on-center magnocellular retinal ganglion cell. (A) Schematic of the receptive field. The cell receives input from a mix of red-, green-, and blue-sensitive cones covering the center of the receptive field, indicated by the white circles. Their inputs add up to provide the "On" center response. These inputs are balanced by inputs from a mix of red-, green-, and blue-sensitive cones—indicated here by the dark circles—covering a ring-shaped surround. These inputs are switched in sign relative to the center and constitute the "Off" surround. (B) Schematic of this cell's response to light. The cell normally fires at some background rate, indicated by the steady, ongoing "spiking." A spot of light shone on the center during the interval indicated by the heavy black line evokes high spiking activity (top row of figure). If the spot is shone on the surround, it reduces or suppresses the ongoing spiking (middle row). The excitation by the center and suppression by the surround largely cancel each other out, so a large spot of light covering the entire receptive field leads to essentially no change in the ongoing firing (bottom row).

the cell remains silent. These cells will form our exemplar when examining how visual processing in the eye matches the structure of the visual world (see figure 6).

To gain some perspective on the function of our eye's receptive fields, it is helpful to take another little detour, this time into the visual system of a species very different from our own: the frog. The frog's eye is built on the same basic plan as ours. It has similar light-focusing optics, rods and cones, and retinal circuits that process the visual input and sends it out via optic nerve fibers emerging from retinal ganglion cells. The processed output is remarkably well tuned for the frog's visual world. This was shown in a classic 1959 paper, "What the Frog's Eye Tells the Frog's Brain," by Jerome Lettvin, Humberto Maturana, Warren McCulloch, and Walter Pitts, then at MIT.[3] The authors found that the electrical responses of individual optic nerve fibers fell into a few major classes. The fastest class of optic nerve fibers responded to a sudden darkening over a large

visual region, exactly the sort of visual stimulus that might signal a loom-
ing bird of prey diving down toward the frog. Another major class of output neurons appeared to be "bug detectors." These nerve fibers responded
best when their receptive fields were presented with dark, bug-sized
patches of just the right size, not too large or too small, with crisp boundaries, preferably moving jerkily. This again makes sense given the frog's
diet of bugs and the fact that a bug has to be moving for a frog to eat it;
frogs will starve if they are surrounded by bugs that are immobilized.
Another class of neurons responded best to extended edges in their receptive fields, likely relevant for the animal jumping from one extended
ledge or leaf to another. It is important that each of these output neurons
responded to its preferred visual shape or pattern independent of overall
brightness. Lettvin and his colleagues found that a given nerve's response
remained vigorous and specific to its preferred stimulus pattern, even as
the illumination level was changed more than three hundred–fold, from
bright incandescent light to something so dim the experimenters could
barely see.

Our ecological needs are quite different from those of the frog. Detecting flying bugs and swooping birds of prey is not central to our lives.
We—and our primate ancestors and relatives—survived by dealing with
much broader classes of objects in our surroundings, such as other animals (Are they dangerous? How fast are they moving?), a fruit (Is it ripe?),
or the branch of a tree (How high is it and how far away?).

But before getting to such complex questions, we need to consider
how our visual system needs to effortlessly parse the visual scene in front
of us into separate objects and distinguish these objects from their backgrounds and from each other. Visual processing in our retina appears to
be suited to precisely that task, in the particular context of the visual world
around us. The essential features of our visual world that let the retina
start parsing it into objects can be appreciated in almost any image, such
as the one in figure 7 of the Hudson River, taken from my office at Columbia Medical School. The image contains extended surfaces whose
visual features change very slowly across space, bounded by visual edges
that typically separate an object from its neighbor or its surround. We see
this in the "natural" portions of the image, with the sky and clouds, and
we see it in the built portions, in the roads, buildings, or cars. We can see
right away that neurons with the on-center magnocellular receptive fields

FIGURE 7. View from my office window. Nearby regions in a scene are likely to share the same brightness (circle B versus circle A); as you move farther away, this likelihood decreases (circle C versus circle A).

described above would be useful. Such neurons would be silent over the uniform regions such as the sky or water or the building wall. They would respond to some degree at the fluffy edge of the cloud. They would respond more vigorously at the crisp edge of the river bank or the edge of the wall, indicating possible object boundaries.

But the match between the properties of visual scenes and the properties of cell responses in the retina is not just qualitative; it is actually quantitatively quite precise. We can measure the spatial distributions of the patterns of light and dark. In a scene, if one spot has a certain lightness, we can ask, "What is the probability of encountering that same lightness as a function of distance away from that spot?" The image can also be analyzed into so-called spatial frequencies—a measure of how quickly visual features change in space. That tells us how much energy in the image is devoted to slowly or moderately or rapidly fluctuating patterns. Such an analysis shows that slowly changing features, such as the sky, dominate. Features like the clouds, the canopy of leaves, and the wall's edge, which show progressively more rapid fluctuations, contribute correspondingly less to the overall energy in the image. Visual scientists have analyzed banks of images in terms of these so-called statistics of the distribution of contrast in the scene. It is striking, and indeed surprising, that almost across the board, images—whether of natural or built environments—have very similar statistics.

It is particularly notable that the magnocellular neuron's response preference is almost the precise opposite of these scene statistics. It's as though that bit of the eye's circuit were built to ignore what is expected in typical scenes and focus on what is unusual or unexpected. The magnocellular neuron is least responsive to slowly varying visual fluctuations, almost exactly to the extent that such slow visual fluctuations dominate visual scenes. Increasingly rapid visual fluctuations—which account for progressively less of the energy in a typical scene—drive the magnocellular neuron progressively more vigorously. And so on up to the most rapidly fluctuating portion, such as the texture of the leaves or the sharp edges of walls. So in a mathematically exact way, the overemphasis on slow fluctuations in the visual world is flattened out and de-emphasized by the magnocellular neuron's responses so as to pick out precisely where things change, where the boundaries lie, and where objects are likely to be defined.[4]

As an added benefit, this leads to a huge image compression in our eyes. The information from the roughly 120 million rods and cones of our retina is compressed one hundred–fold into the relatively wispy one million fibers of the optic nerve. This also allowed our eyes to evolve into eyeballs that can rotate effortlessly in their sockets as we look around, something we would never have been able to do if each of the 120 million rods and cones were wired up individually to the brain because such wiring would create a thick, cumbersome tether.[5]

Finally, not only is it important for us to quickly pick out objects, but also we should be able to do so independently of how brightly or dimly they are lit. We can do that because the retinal circuit has evolved ways of adapting rapidly to local light intensity so as to maintain a consistent response to visual patterns independent of intensity.[6] This allows our visual system to operate more or less equally well over a ten billion–fold range of light intensities. To place that in context, a jpeg image carries only 8 bits, or 256 light intensities. The best professional digital cameras of 2017 claimed dynamic ranges of about 14 stops, which translates to 2^{14}—that is, 16,384 light intensities. This means our visual system operates effortlessly over light levels a million times brighter than the saturation level, or a million times darker than the darkest shade that can be detected by the best digital cameras. That is just as well. It allowed our forebears to be wary of that lion or chase that antelope whether in bright sunlight or in approaching dusk. And it allows us to similarly avoid that moving car and step effortlessly on to the curb, however bright or dim the light. This adaption to scene brightness also makes us rather poor at quantifying ambient brightness levels beyond just the sense that "it is bright" or "it is dark." But that does not matter since it is object size, shape, and identity, rather than ambient light level, that is most important to our survival.

This fine matching of vision with the statistics of natural scenes to pick out features that are important does not stop at the eye. We also see it in later stages of visual processing in the brain. Circuits in the brain are more complex, so the evolutionary principles are harder to state and test mathematically. And evolution is difficult to undo or manipulate, so one cannot "prove" that a particular step in the visual processing evolved to match our natural surroundings. But the match is tantalizingly robust and specific to our needs. We would not survive our complex world if our eyes were specialized for detecting moving bugs and diving birds, while for the frog, spending that extra second to identify the diving bird using the more general-purpose human visual system could mean the difference between life and death. It is tough to resist the conclusion that vision evolved for our survival, in the particular world into which we were born. Not just the eye but all the components of vision know what is good for us.

NOTES

1. C. D. Gilbert, "The Constructive Nature of Visual Processing," in *Principles of Neural Science*, ed. E. R. Kandel, J. H. Schwartz, T. M. Jessell, S. A. Siegelbaum, and A. J. Hudspeth (New York: McGraw Hill, 2013), p. 556.

2. M. Meister, "Low Level Visual Processing: The Retina," in Kandel et al., eds., *Principles of Neural Science*, p. 577.

3. J. Y. Lettvin, H. R. Maturana, W. S. McCulloch, and W. H. Pitts, "What the Frog's Eye Tells the Frog's Brain," *Proceedings of the Institute of Radio Engineers* 47 (1959): 1940–1951.

4. J. J. Atick, "Could Information Theory Provide an Ecological Theory of Sensory Processing?" *Network* 22 (1992): 4–44.

5. H. B. Barlow, *Physical and Biological Processing of Images* (Berlin: Springer-Verlag, 1983).

6. Adaptation to light intensity and contrast are very important aspects of vision.

We are all familiar with slow adaptation to large, overall changes in ambient light, such as walking in from bright sunlight into a dark room and only gradually starting to make out the objects in the room. But adaptation also occurs continually, and much more rapidly, over a couple of hundred milliseconds. In normal vision our eyes are continually making rapid movements, known as saccades, that bring different bits of the scene on to our fovea, where we have the best spatial resolution. This happens a few times a second. A typical sunlit scene contains about a ten thousand–fold variation in local light intensity, from brightly sunlit to shaded patches. So saccades would subject any local bit of the retina—the fovea, say—to equally large fluctuations. The retinal circuit adapts over the 200–400 milliseconds that the eye dwells at any location. Photoreceptors adapt for relatively large changes in local light intensity. Other elements of the circuit—including the intermediate bipolar cells linking photoreceptors to ganglion cells and the lateral feedback cells know as horizontal cells and amacrine cells—adapt across smaller changes of light intensity. This adaptation allows the retina to encode the ten thousand–fold variation in local intensities and contrasts into the much narrower range of information the optic nerve can carry—a maximum of about one hundred distinct levels, given the typical peak firing rate of about two hundred spikes per second. This form of adaptation happens on line, continually, while we look around. In addition, there are slower levels of adaptation that take a few seconds to a few minutes. This allows the circuit to go from bright sunlight (10^6 cd / m^2) to sunset or sunrise (about 10 cd / m^2) with little change in our ability to discriminate shapes or colors. F. Rieke and M. E. Rudd, "The Challenges Natural Images Pose for Visual Adaptation," *Neuron* 64 (2009): 605–616; J. B. Demb, "Functional Circuitry of Visual Adaptation in the Retina," *Journal of Physiology* 586 (2008): 4377–4384.

You Have a Superpower—It's Called Vision

Charles E. Connor

FOR THOSE GIFTED WITH NORMAL SIGHT, vision is a superpower. At a glance, you can tell where you are, what is around you, what just happened, and what is about to happen. You effortlessly perceive the precise three-dimensional structure of objects in your environment at distances ranging from millimeters to miles. You know what they are called, how valuable they are, how old or new, fresh or rotten, strong or weak. You intuit their material, mechanical, and energetic properties, allowing you to anticipate and alter physical events with deadly accuracy. You effectively read the minds of other humans and animals based on tiny variations in facial configuration and body pose. A picture is worth many times a thousand words to you.

All this information seems to exist outside you, immediately and effortlessly available. Understanding what you see seems trivial—you only have to *look* at it! We are so good at vision that we don't even recognize it as an ability. Moreover, we live in a world where most people are gifted with the same superpower. Vision is ordinary. Other people see the same things you do—no big deal.

But let me try to convince you of how *extra*-ordinary vision is. Imagine that you travel to a world of sentient, humanoid creatures without vision; call them Glorbons. These creatures possess our other sensory and motor abilities, as well as comparable intelligence. They understand

and interact with their environment based on touch, hearing, taste, and smell. Think how magical your visual abilities would seem in such a world. Your immediate, accurate descriptions of all the things in their houses and cities, regardless of distance, things they know only through laborious close-up exploration, seem supernatural. You have an oracular ability to forecast physical events like oncoming storms. Your ability to recognize individuals from afar and intuit their health, mood, age, and intentions before they even speak resembles witchcraft. You are an action hero: you can anticipate stealthy crimes and sneak attacks, and your accuracy with distance weapons is uncanny. To the Glorbons, you possess inexplicable, godlike knowledge and power.

You try to demystify this ability to the Glorbons by teaching them about how vision works. Drawing on your dimly remembered high school physics, you explain that vision is based on photons—particles, or maybe waves, that weigh absolutely nothing and travel through space at about a billion feet per second. To the Glorbons, of course, photons sound no more convincing than midi-chlorians do to us.[1] But you forge ahead, telling them how the paths of photons can be bent by the lens in your eye to form an image, a two-dimensional map in which each point on the retina, the layer of light-sensitive cells at the back of the eye, receives photons only from a corresponding direction in space. "Okay," respond the Glorbons. "What do you do then?" "Well," you reply, "then you just see what is in the image." But the Glorbons don't understand what "seeing" means. And as you try to explain, you begin to realize that you don't understand seeing either. It is something you do all the time, but you don't think about doing it, and you don't know how it happens. There *is* something mysterious about vision, even to you.

Promising the Glorbons that you will return, you travel back to earth, where you are sure that scientists will be able to explain seeing. First, based on what you've been reading lately about convolutional neural networks[2] (computer simulations of densely connected neural hierarchies), you consult an expert on computer vision, Professor Y. You describe your difficulty in explaining to the Glorbons. He commiserates and tells you that vision has turned out to be one of the most difficult problems in computer science, in spite of early confidence. He shows you a project outline from MIT in the 1970s, predicting that computer algorithms for recognizing objects in photographs could be designed by summer students.[3] But now, he says, nearly fifty years later, it remains a largely unsolved problem.

But surely, you say, convolutional networks have finally replicated human visual abilities? Professor Y smiles and takes you to his lab to see AlexNet, a standard deep computer network, in action.[4] He brings up a picture of a penguin standing in a snowy field, with some other penguins and mountains in the background. The network draws a box around the penguin and labels it with a 70 percent probability that it is a bird. Pretty good, you think, though you personally would have given it a 100 percent chance of being a *penguin*—it's really unmistakable, with that black beak; beady black eyes squished deep into either side of its head, which is tilted quizzically to one side; its downy stomach, looking warm in spite of the snow; its silly looking short wings that are really superb flippers; and its unmistakable pear-shaped body puddling out over and nearly obscuring its black webbed feet.

Next, Professor Y adds a computer monitor into the picture, right next to the penguin. You start to laugh at the Pythonesque incongruity, but then you notice that the network has now labeled the penguin with a 70 percent probability of being a *person*.[5] How can that be, you say—it's obviously a penguin; it doesn't look like a person at all. Professor Y explains that this is how the most successful object identification programs work, using statistical inference on cues from the entire image to make a best guess. But, you say, what about the beak, the down, the wings . . . and then you realize that AlexNet doesn't have all of this overly complete information about the physical reality of the penguin that makes it impossible to mistake for anything else. Instead, AlexNet learns the simplest, most reliable cues for object identification, its only task. Your exuberant perception of penguin-information is way more than you need for coarse identification under normal circumstances. (But it comes in pretty handy if you want to *understand* penguins, their physical nature, their evolutionary adaptations, their behavior, etc.)

Clearly, you think on your way out, seeing is much more than object identification. You need someone who understands *human* vision, so you consult a visual neuroscientist, Professor Z. You tell her about your interest in learning how human vision works—the rich complexity, detail, and meaning of it, not just identification. She has a wealth of information to share about the last sixty years of research on high-level visual processing.

First, she tells you about anatomy—how perhaps half of the cerebral cortex in primates (including humans) is involved in vision in some way. The primary visual cortex, at the back of the brain, which receives visual information detected and processed by the retina (see essay by Aniruddha Das in this volume) and then further transformed by an intermediary structure called the thalamus, is just the start. From the primary visual cortex, information is distributed through dozens of other brain regions involved in various aspects of visual experience.[6] There are specific regions devoted to objects, scenes, faces, and bodies and to movement and color. Action-related areas use vision to decide what to look at next, how to navigate through the world, how to reach out and grab objects. Regions associated with emotion and appetite use visual information to evaluate how valuable or dangerous things are. In the prefrontal cortex, visual information is stored in short-term memory, for seconds. In the temporal cortex, visual information is stored in long-term memory, for lifetimes.

Now, this is interesting. No wonder visual experience is so rich and information-dense, with half the cerebral cortex working on it every waking minute. (Professor Z assures you that the well-known phrase "You only use 10 percent of your brain" is a canard.) But, you say, that is all about *where* vision happens. What I really want to know is *how* it works. "Ah, the $64,000 question. . . . It's complicated. We've had a few successes; we probably know the most about visual motion perception—how groups of neurons signal direction and speed."[7] Oh, you say, for example, how we appreciate the ballet? "Not quite; more like how we decide whether a field of dots is moving left or right." What about fine art—how do I get so much out of staring at a Vermeer? "We know bits and pieces— neural signals for edges,[8] colors,[9] three-dimensional surfaces[10]—but how it all gets put together? You'd need to be God, able to measure every neuron in the brain and understand what they are all saying to each other."[11]

Dejected, you walk away, deciding that visual neuroscience is full of profound questions but disappointing answers. Vision is a superpower, and it remains mysterious, for now. The Glorbons will have to wait.

NOTES

1. Midi-chlorians are the microscopic, intracellular, intelligent symbiotes who mediate the Force according to "The Phantom Menace" (*Star Wars*, episode 1).

2. Y. LeCun, Y. Bengio, and G. Hinton, "Deep Learning," *Nature* 521 (2015): 436–444.

3. S. Seymour Papert, "The Summer Vision Project," MIT AI Memo 100, Massachusetts Institute of Technology, Project Mac, 1966.

4. A. Krizhevsky, I. Sutskever, and G. E. Hinton, "Imagenet Classification with Deep Convolutional Neural Networks," in *Advances in Neural Information Processing Systems* (Cambridge, MA: MIT Press, 2012), pp. 1097–1105.

5. A. Yuille, "Learning in the Visual Brain: Towards Understanding Deep Neural Networks," Biennial Science of Learning Symposium, Johns Hopkins Science of Learning Institute, 2016; http://scienceoflearning.jhu.edu/events/biennial -science-of-learning-symposium-presentations.

6. D. C. Van Essen, M. F. Glasser, D. L. Dierker, J. Harwell, and T. Coalson, "Parcellations and Hemispheric Asymmetries of Human Cerebral Cortex Analyzed on Surface-Based Atlases," *Cerebral Cortex* 22 (2012): 2241–2262; D. J. Kravitz, K. S. Saleem, C. I. Baker, L. G. Ungerleider, and M. Mishkin, "The Ventral Visual Pathway: An Expanded Neural Framework for the Processing of Object Quality," *Trends in Cognitive Sciences* 17 (2013): 26–49; D. J. Kravitz, K. S. Saleem, C. I. Baker, and M. Mishkin, "A New Neural Framework for Visuospatial Processing," *Nature Reviews Neuroscience* 12 (2011): 217–230.

7. A. J. Parker and W. T. Newsome, "Sense and the Single Neuron: Probing the Physiology of Perception," *Annual Review of Neuroscience* 21 (1998): 227–277.

8. D. H. Hubel and T. N. Wiesel, "Receptive Fields and Functional Architecture of Monkey Striate Cortex," *Journal of Physiology* 195 (1968): 215–243.

9. M. S. Livingstone and D. H. Hubel, "Anatomy and Physiology of a Color System in the Primate Visual Cortex," *Journal of Neuroscience* 4 (1984): 309–356.

10. Y. Yamane, E. T. Carlson, K. C. Bowman, Z. Wang, and C. E. Connor, "A Neural Code for Three-Dimensional Object Shape in Macaque Inferotemporal Cortex," *Nature Neuroscience* 11 (2008): 1352–1360.

11. "Fact Sheet: BRAIN Initiative," White House Office of the Press Secretary, April 2, 2013.

The Sense of Taste Encompasses Two Roles

CONSCIOUS TASTE PERCEPTION AND
SUBCONSCIOUS METABOLIC RESPONSES

Paul A. S. Breslin

MOST OF US KNOW the sense of taste as an oral component of food flavor or perhaps the bitterness of medicines. But most people probably do not consider what our sense of taste does for us. That is, how is taste useful? In our daily lives taste operates as part of a set of highly integrated, specialized sensory systems we call "the chemical senses," comprised of taste, smell, and the skin responses to chemicals, such as the burn of chili peppers.[1] To understand their importance, I think it is instructive to consider their roles in human survival and evolution. One evolutionary perspective on biology, which I favor, is that our anatomy, physiology, and psychology are intimately tied to helping us eat food and have sex. Our search for food has shaped us (and all animals) significantly because it is necessary to keep us going into tomorrow by providing energy and nutrients. Our search for sex has shaped us because it is necessary to keep our species going into the distant future by producing offspring. The chemical senses are essential to feeding and mating in most animals ranging from ants to dogs. It is easy to imagine the importance of a dog's sense of smell for finding food and mates, but the role of taste in these processes is somewhat more obscure. This is, in part, because olfaction is a distal sense that can guide from afar, as vision and hearing do, whereas taste, at least for nonaquatic animals, is a more intimate, proximal sense that requires contact with the food or partner in question. For many invertebrates, such as insects, taste is necessary for both eating and mating.[2] But for mammals the function of taste in sex is not yet clear, although licking mates' mouths and genitals is common to a wide range of species. In contrast, the necessity of taste for eating is more obvious. Here, I will focus on the human (and mammalian) sense of taste to illustrate its relationship to feeding and its dual functions therein.

The sense of taste has two fundamental roles in relation to our eating: to help us identify and recognize foods to establish their palatability—that is, their "yumminess and eatability"—and to prime digestive and metabolic activities that will optimize the efficiency of how nutrients from food are processed and used by the body. The first function generates conscious sensations, so we are familiar with it, and the second function proceeds subconsciously. Why is this second function of metabolic priming important? To use an analogy, imagine what air travel would be like if every time a plane landed at an airport, it was a big surprise. Figuring out the flow of traffic in the air and on the tarmac would be uncertain and dangerous. Determining which gate to put the plane at would be slow, and finding the gate personnel to show up and move the disembarking equipment and passengers would be highly difficult. Furthermore, unloading luggage and determining whether it was staying at that location or continuing on to a new destination would be chaotic, as would be refueling the plane, cleaning it, stocking it with food, and so forth. What makes modern air traffic possible is the anticipation by the airport of all of these needs and the preparedness of the equipment, personnel, and supplies for action. So what happens when we eat if we do not anticipate incoming nutrients? The point is made eloquently by Steve Woods, who wrote, "The act of eating, although necessary for the provision of energy, is a particularly disruptive event in a homeostatic sense."[3]

"Homeostasis" is an idea first raised by the physiologist Claude Bernard (1813–1878), who wrote, "The constancy of the internal environment is the condition for a free and independent life."[4] This idea was adapted by Walter Cannon, who, in 1932, coined the term "homeostasis" in his book *The Wisdom of the Body*.[5] Cannon identified principles for an integrated system that tends to maintain relatively constant internal environments, including the idea that our regulatory systems are made of multiple components that are organized to work in concert to increase or decrease concentrations of critical nutrients and metabolic factors in the blood to achieve desirable levels. Such a constant environment enables our ability to migrate, run, hunt, work, or sleep independently of whether we happen to be eating, digesting, or fasting.[6] Taste is an essential warning circuit that informs these coordinated homeostatic systems of what is about to enter the body, like a nutritional air-traffic control tower. This includes warning that macronutrients, micronutrients, and water, as well as toxins and poisons that interfere with normal metabolic processes, are about to be consumed and absorbed via the gastrointestinal tract.[7] Anticipatory reactions to these foods and nutrients can include gastric churning and intestinal rhythmic contractions; exocrine secretions, such as saliva into the mouth or gastric acid into the stomach; endocrine secretions, like early insulin release into blood; and molecular responses, such as the increase in the proteins that shuttle nutrients across the intestine and into the blood. The absence of anticipation of nutrients makes the body appear a bit like the airport that is not expecting incoming airplanes—nothing works right. For example, Ivan Pavlov observed that placing lumps of meat directly into dogs' stomachs resulted in no digestion. However, when he simultaneously gave a dusting of dried meat powder onto the dogs' tongues, then the lump of meat placed into the stomach was completely digested.[8] In short, his Nobel Prize–winning life's work can be summarized as follows: for us to be healthy, proper digestion requires that incoming food and nutrients be anticipated and that the body be well prepared.

How does the sense of taste respond to nutrients and toxins? Taste is normally stimulated when food enters the mouth and acts upon taste receptor cells. These are found within microscopic, rosebud-shaped clusters called taste buds, located on the tongue, soft palate (back part of the roof of the mouth), and pharynx (back of the throat). On these receptor cells are specialized proteins (receptors) that allow these electrically active cells to generate currents when stimulated by chemicals in the mouth. We generally do not have taste receptor cells or taste buds under our tongue, on our lips, on the inside of our cheeks, or on our hard palate (the front portion of the roof of the mouth). We do, however, have taste receptors throughout our digestive tract and in all of our major metabolic organs, with which we monitor blood nutrient levels.[9] Some have argued that the front of the mouth and the back of the mouth serve different purposes—with the tip of the tongue used to probe and identify what we are tasting/sampling and the back of the tongue, soft palate, and pharynx used more to help determine whether the food should be swallowed or spat out. The receptors in the back of the mouth contribute to the go/no go decision of eating. This latter component is of ultimate importance not only because we need nutrients to live (go), but also because, and even more important, if we eat poison, one meal could kill or sicken us (no go). It is important that receptors in the pharynx are stimulated only after swallowing; that is why beer tasters are correct in saying that you need to swallow beer to evaluate its flavor fully.[10] So how could pharyngeal receptors help us decide whether or not to swallow? First, the decision of whether to swallow or not to swallow is an ongoing process during feeding rather than a "once and done" decision. Second, pharyngeal taste is a final safety check for what we are about to absorb after swallowing in case we need to abort the feeding by trapping food in the stomach and regurgitating it. This is one reason that stimulation of the back of the throat can induce gagging, nausea, and vomiting. Overall, the taste sensory system of the oral cavity communicates with the brain and the metabolic organs neurally, as well as through the secretion of hormones into the circulating blood.[11]

The conscious perception of a taste can be referred to as a quale, a singular sensation, such as when licking table salt. Yet the taste qualia we experience are actually comprised of separate attributes. Taste quale may be subdivided into the components of sensory quality (sweet, sour, salty, savory/umami, bitter, and perhaps a few others such as water taste, malty taste, or mineral taste); intensity (weak to strong); location (such as a bitter taste on the tip of the tongue or a bitter taste on the back of the tongue); and temporal dynamics (a short-lived taste or a lingering aftertaste). These features of taste are typically combined in the brain with a food's other oral sensory and olfactory properties to create its flavor, to help us identify and recognize the food, to help reassure us that what we are experiencing is edible, and to create an association with how we feel after eating so that we can recognize the food at our next encounter and remember whether it was satisfying or made us sick.

Because the taste system is connected to the brain and the metabolic organs, such as the pancreas and the liver, when foods are in the mouth, we are able to evaluate them as either rewarding (i.e., experienced as yummy) or punishing (i.e., experienced as yucky or nauseating). High-caloric, nutrient-dense foods are rewarding and lead to acceptance of the food, whereas foods laden with toxins and poisons are punishing and lead to rejection. Consciously, a food may be identified as containing nutrients from its taste qualities, particularly when it is sweet, savory, sour, or salty tasting, and these taste systems trigger reward centers of the brain by signaling that beneficial nutrients are coming in. Similarly, toxins will often give rise to bitter taste, which we consciously perceive. The bitter taste system, if stimulated strongly, is clearly punishing, although we generally tolerate weak bitterness. The positive behavioral reflexes of licking and swallowing for rewarding stimuli, as well as the rejection reflexes of drooling, gaping, head shaking, and shuddering for punishing stimuli, are automatic reflex responses. They are controlled by subconscious brainstem sensory-motor reflex arcs, much as the removal of a hand from a hot stove is driven subconsciously via the spinal column.

Other subconscious responses to oral stimuli can directly influence blood hormone levels, digestion, and blood nutrient levels. This means that the taste system directly informs and influences metabolic processes. There are several rather remarkable examples of taste stimulation accomplishing this. For instance, if we return to our analogy of airline preparedness, we know that clearing glucose from the blood is necessary for life and requires insulin when we are not exercising, so the taste system gets this process going as soon as we taste carbohydrates (sugars and starches). Rinsing with glucose orally and spitting it out (that is, even without swallowing it) can cause an anticipatory release of insulin from the pancreas. This release is very small but has a large effect on lowering blood glucose during and after eating. Surprisingly, this occurs not only for sugar, but also for oral starch, such as we may find in bread, potatoes, or pasta.[12] But it is not clear yet that we have starch taste receptors. We do, however, have the enzyme salivary amylase in our mouths that digests starch sufficiently to become a taste stimulus, which in turn can elicit anticipatory insulin release. Another oral response to carbohydrates involves performance enhancement. It is well known that human exercise performance is boosted when athletes eat glucose during competition. But the same effect can be found when glucose (or starch) is rinsed and spat out. Yet rinsing and spitting out the noncaloric sweetener saccharin has no impact on performance. Thus it appears that the taste system identifies which stimuli are metabolizable carbohydrates and then influences our performance through means that are not yet understood.

We also find anticipatory effects from another macronutrient, oral fat. People who chew and spit out real cream cheese on a cracker versus nonfat cream cheese (a.k.a. fake cream cheese) on a cracker have relatively large elevations in triglycerides measured in the blood plasma for hours after.[13] This study suggests that whole body lipid metabolism may be regulated by oral sensations of fats. The effects of taste on macronutrient metabolism have been applied to help premature and sick infants. Babies who must receive all their nutrition by tube feeding or intravenously learn to feed orally faster and have shorter hospital stays if they can suck on a sweetened, flavored pacifier at the same time that nutrients are placed into their stomach or vein.[14] This effect suggests that non-nutritive sucking of a sweetened pacifier alters how the nutrients placed into the stomach or a vein are handled and utilized by the baby's body.

In addition to macronutrient responses, there is further evidence that taste and other oral sensations cause anticipatory responses to maintain salt and water balance in the body. We know that drinking water quenches our thirst. But our thirst is quenched and drinking ceases long before the water in our bellies is absorbed and can rehydrate us. This means that oral sensations from water, possibly water taste, are used as cues to anticipate how much water will be absorbed. Even simply rinsing and spitting out cold water helps relieve thirst. Since our bodies must maintain relatively constant concentrations of salts in our blood, ingestion of salt and water typically balance each other. Therefore, we also have anticipatory reflexes to the mere taste of salt. In rodents, salt water in the mouth causes metabolic reflexes of the kidney. A drop of concentrated salt water placed on a rat's tongue causes immediate decreases in the production of urine in order to retain water and maintain salt and water balance of body fluids.[15]

We also anticipate the ingestion of toxins. Many toxins taste bitter, and toxins or poisons also make us feel sick and nauseated. It is interesting that strong bitter tastes are sufficient to induce nausea and also to cause the stomach to stop its normal rhythmic churning.[16] So we anticipate incoming toxins when we taste strong bitter tastes by experiencing the punishing effects of nausea, readying us to regurgitate, and by trapping the contents of the stomach, preventing us from absorbing what we ingested. These preparations for incoming toxins can be lifesaving.

I believe that nearly all taste stimuli both elicit a conscious taste perception and stimulate a subconscious metabolic reflex. This is true of both nutritious and toxic stimuli. The dual nature of taste makes it arguably the most indispensable sensory system humans possess. Taste resists decline as a function of aging more than the other senses, perhaps because of its necessity for life.[17] In contrast, humans learn to cope with sensory deficits such as an inability to smell (anosmia), blindness, and deafness. But in the rare instances that humans lose their sense of taste, they do not eat much, they lose weight, and their nutritional state falters. One of the more readily observed ways people lose their sense of taste is following radiotherapy treatment of head and neck cancers. Taste loss with oral radiotherapy is common, and the impact on patients can be shocking. I leave you with a quotation from E. M. MacCarthy-Leventhal, a medical doctor who underwent radiotherapy for her head and neck cancer and lost her sense of taste as a result:

> What is it like to lose your sense of taste? To know that the most luscious fruit is a cinder, and its juice flavored with copper and bicarbonate, or that a Whitstable oyster is no more appetizing than a slug? If, by a might of effort, these "cinders" are forced down with copious fluid, the consequences are acute indigestion and vomiting. The patient is not hungry anyway, and it is easier to starve.[18]

NOTES

1. The sensations on the skin or mucous membranes that include the burn of chili peppers, the coolness of mint, or the eye-watering properties of onions are sometimes called chemical somesthesis or "chemesthesis" for short. B. G. Green, "Measurement of Sensory Irritation of the Skin," *American Journal of Contact Dermatitis* 11 (2000): 170–180.
2. H. Matsunami and H. Amrein, "Taste and Pheromone Perception in Mammals and Flies," *Genome Biology* 4 (2003): 220–229.
3. S. C. Woods, "The Eating Paradox: How We Tolerate Food," *Psychological Review* 98 (1991): 488–505.
4. Quoted in C. G. Gross, "Claude Bernard and the Constancy of the Internal Environment," *Neuroscientist* 4 (1998): 380–385.

5. W. B. Cannon, *The Wisdom of the Body* (New York: W. W. Norton, 1932).

6. Gross, "Claude Bernard and the Constancy of the Internal Environment."

7. Nutrients are made up of macronutrients, micronutrients, and water. The term "macronutrients" refers to our need for and ingestion of carbohydrates, proteins, and fats in relatively high levels. The term "micronutrients" refers to our need for and ingestion of minerals and vitamins in relatively low levels. All of these nutrients, including water, are generally monitored, and levels are regulated to some degree. Note that some are regulated very precisely, such as certain vitamin levels, whereas others are allowed to vary much more highly, such as plasma fat levels.

8. I. P. Pavlov, *The Work of the Digestive Glands* (London: C. Griffin, 1902).

9. Taste receptors are found not only in the mouth and throat, where they can generate conscious taste perception, but they are also located on the inner surface of the entire gastrointestinal tract, where they unconsciously alter metabolism rather than generate conscious taste sensations (thankfully). Taste receptors are also expressed in organs, where nutrients stimulate them post-absorptively, although in these locations they do not form taste buds. These include most major metabolic regulatory organs and tissues including the liver, the pancreas, fat cells (adipocytes), the thyroid, and the brain. A. Laffitte, F. Neiers, and L. Briand, "Functional Roles of the Sweet Taste Receptor in Oral and Extraoral Tissues," *Current Opinion in Clinical Nutrition and Metabolic Care* 17 (2014): 379–385.

10. Wine tasters are noted for swishing and spitting wine rather than swallowing wine when sampling. They do so, in part, to avoid becoming intoxicated when sampling multiple wines. But the same logic applies to wine as to beer, and more complete tasting would come from swallowing the wine. That said, one important difference between beer and wine is the hop vine cone flower extracts that are added to beer. In these hop flower extracts are a class of bitter-tasting compounds called the iso-alpha-acids that are particularly bitter tasting in the throat.

11. Oral taste buds, which are multi-cellular micro-organs, can also function as endocrine organs secreting measurable levels of metabolic regulatory hormones into the blood. Z. Kokrashvili, K. K. Yee, E. Ilegems, K. Iwatsuki, Y. Li, B. Mosinger, and R. F. Margolskee, "Endocrine Taste Cells," *British Journal of Nutrition* 111 (2014): S23–S29.

12. A. L. Mandel and P. A. Breslin, "High Endogenous Salivary Amylase Activity Is Associated with Improved Glycemic Homeostasis Following Starch Ingestion in Adults," *Journal of Nutrition* 142 (2012): 853–858.

13. R. D. Mattes: "Oral Fat Exposure Alters Postprandial Lipid Metabolism in Humans," *American Journal of Clinical Nutrition* 63 (1996): 911–917, and "Oral Fat Exposure Increases the First Phase Triacylglycerol Concentration Due to Release of Stored Lipid in Humans," *Journal of Nutrition* 132 (2002): 3656–3662.

14. J. P. Foster, K. Psaila, and T. Patterson, "Non-Nutritive Sucking for Increasing

Physiologic Stability and Nutrition in Preterm Infants," *Cochrane Database of Systematic Reviews* 10 (2016): CD001071.

15. S. Nicolaidis, "Effects on Diuresis of the Stimulation of Buccal and Gastric Afferent Fibers by Water and Saline Solutions," *Journal de Physiologie* (Paris) 55 (1963): 309–310.

16. C. Peyrot des Gachons, G. K. Beauchamp, R. M. Stern, K. L. Koch, and P. A. Breslin, "Bitter Taste Induces Nausea," *Current Biology* 21 (2011): R247–248.

17. J. C. Stevens, L. M. Bartoshuk, and W. S. Cain, "Chemical Senses and Aging: Taste versus Smell," *Chemical Senses* 9 (1984): 167–179; L. P. Schumm, M. McClintock, S. Williams, S. Leitsch, J. Lundstrom, T. Hummel, and S. T. Lindau, "Assessment of Sensory Function in the National Social Life, Health, and Aging Project," *Journal of Gerontology: Social Sciences* 64B (2009): i76–i85.

18. E. M. MacCarthy-Leventhal, "Post-Radiation Mouth Blindness," *Lancet* 2 (1959): 1138–1139.

It Takes an Ensemble of Strangely Shaped Nerve Endings to Build a Touch

David D. Ginty

SYMPHONIC ENSEMBLES of neural impulses flow from our skin to our brains to convey tactile encounters with the physical world. What are the instruments of these neural ensembles? How does one composition of impulses signify a mother's nurturing touch and others a menacing bug, a summer's breeze, a tickle, a complex pattern of Braille dots, or raindrops? Our ability to perceive and respond to the physical world is rooted in the cadre of intricate nerve cell endings in our skin, the largest and least understood of our sensory organs.

Nerve cells that respond to innocuous (nonpainful) tactile stimuli acting on the skin are called touch receptors. The cell bodies of touch receptor neurons form clusters called ganglia adjacent to the spinal cord, and they extend one long axon to the skin and another that enters into the spinal cord (or brainstem). Excitation of touch receptor endings in the skin is the first step leading to the perception of touch, so these touch receptor neurons serve as neural portals to the vast physical world. Although no one knows for sure how many different types of innocuous touch receptors we mammals have, detailed physiological measurements yield an estimate of about seven, with perhaps a few types remaining to be discovered.[1] Upon skin stimulation and excitation of a touch receptor, electrical impulses flow along its axon from the skin into the spinal cord, where they are combined with impulses arriving from other touch receptor types. From there, second-order neurons convey integrated touch receptor information to higher brain regions (called the thalamus, midbrain, and somatosensory cortex) for interpretation and action—a topic for another day. In this way, our perception of a touch begins with activation of physiologically distinct classes of touch receptors whose sensory endings are embedded within our skin.

Distinguishing features of the seven or so touch receptor types include the speed at which their impulses are propagated from the skin to the spinal cord (ranging from 2 to 200 miles per hour), their rate of adaptation to a sustained touch (some touch receptors continue to fire spikes throughout the touch while others stop firing soon after the touch begins), their ending shapes and locations within the skin, and, most important, their sensitivity to different types of mechanical forces impinging upon the skin and its appendages (e.g., the outermost layer of mostly dead skin cells called the epidermis, as well as hairs, whiskers, and fingernails). Indeed, just as the light-detecting cells of the retina have different sensitivities to light of different colors, touch receptor endings in the skin have distinct but overlapping sensitivities to different mechanical stimuli. One touch receptor, for example, responds strongly to hair deflection, whereas another responds best to high-frequency vibration of the skin.

We can distinguish among an enormous number of different mechanical stimuli; touch is not simply a "yes or no" sensation. We can tell leather from wool, an oily surface from sandpaper, a caress from a poke, and a tickle from a pinch. Indeed, we can identify pointy objects of various degrees of sharpness; skin stretch in response to different body movements or external forces; strokes of the skin with different speeds or with various degrees of pressure; deflections of, or tugging on, hair follicles; vibrating objects acting nearby or directly touching us; and the running of our fingers across objects of different shape, smoothness, contour, roughness, compliance, etc. The point is that our seven or more touch receptor types, with their distinctive sensitivities and impulse firing patterns, collectively enable a virtually limitless number of impulse ensembles that flow from the skin to the central nervous system, thereby informing our brains of a wide range of physical encounters.

What features of touch receptor types underlie their characteristic sensitivities to different mechanical forces? There are two general thoughts about these features. One idea is that each touch receptor type is intrinsically different from the others. That is, each type is genetically determined, through developmental mechanisms, to express distinct protein-based micromachines, which convert mechanical stimuli into electrical signals, and each expresses a unique constellation of ion channel proteins that govern impulse firing rate and pattern. In other words, intrinsic differences in gene expression programs could, in principle, enable touch receptor subtypes to respond to only certain types of mechanical stimuli, send impulses from the skin to the central nervous system at a particular speed, and adapt either quickly or slowly to a sustained stimulus. We know for certain that differences in intrinsic properties of touch receptor types do exist, but the extent to which such differences account for the unique sensitivities of the seven touch receptor types is not known. The second idea is that the touch receptors are distinguished by the uniqueness of their endings within the skin; distinctly shaped endings of touch receptor subtypes endow each with its characteristic sensitivity to different mechanical stimuli.[2] The answer appears to be that both types of difference are important: distinctive electrical properties and highly specialized endings in the skin collaborate to endow each of the seven or so fundamental neural instruments in our touch receptor orchestra with its characteristic combination of sensitivity and impulse-firing pattern.

Francis Crick, co-discoverer of the structure of DNA, arguably the greatest discovery in biology in the twentieth century, famously said, "If you want to understand function, study structure." The early anatomists examined the structure of sensory endings in the skin in their entirety. They saw a vast thicket of neural endings of different shapes and sizes that were seemingly haphazard in their arrangement. This view is not terribly informative for our purpose of understanding the uniqueness of touch receptor types. If, on the other hand, one observes the cutaneous endings of each of the seven touch receptor types, one by one, using modern approaches for genetic labeling of individual cells with fluorescent proteins, then unique, iterative, and logical organizational patterns of each touch receptor ending type are wondrously revealed.[3] Indeed, each touch receptor type adorns spectacularly intricate patterns of cutaneous endings, each one as unique as it is informative. Let's consider three of these.

Mammals have highly sensitive touch receptors that report on bending of hairs. Many mammals also have probing whiskers that extend great distances from the face, acting like curb feelers enabling navigation in the dark. Or sweeping furry tails, strategically placing highly mobile probes equipped with thousands of hairs, survey the physical world from behind. Although we humans are not typically considered tactile foraging mammals—we rarely navigate dark, narrow tunnels—we can nonetheless certainly appreciate the benefits of probing, sensitive whiskers and a furry tail backup detector. Nevertheless, all mammals that have been studied, including us, have touch receptors that are highly sensitive to deflection of body hairs; one of these is called the Aβ RA-LTMR. (It is the technical name given by aficionados; "Aβ" means its electrical impulse propagates rapidly to the central nervous system; "RA" refers to the fact that it is rapidly adapting to indentation of the skin—it fires an impulse during onset and offset of skin indentation; "LTMR" means it is a Low-Threshold MechanoReceptor.) Remarkably, the slightest bending of a single hair can trigger an Aβ RA-LTMR to send electrical impulses from the skin to the spinal cord and ultimately the brain. An explanation for this remarkable sensitivity to hair bending is made readily apparent upon examining the structure of Aβ RA-LTMR endings in hairy skin. Each Aβ RA-LTMR has a single axon that extends into the skin and wraps around hair follicles beneath the skin's surface. Close inspection of these endings reveals a remarkably intimate relationship with hair follicles. The main process penetrates the skin, encircles a hair follicle and elaborates 10–20 pitchforklike smaller processes, called lanceolate endings, which extend along the hair follicle surface. High-magnification images of these longitudinal lanceolate processes with an electron microscope reveal their intimate embrace with the hair follicle's outer layer of ensheathing cells, called epithelial cells.[4] Eureka! This arrangement can explain why Aβ RA-LTMRs are triggered to fire an impulse during hair bending! When a hair is bent, minute displacement of the hair follicle relative to the closely abutted Aβ RA-LTMR longitudinal lanceolate endings causes tugging on the lanceolate ending membrane resulting in an opening of mechanically gated ion channels and thus electrical excitation of the lanceolate endings. Weak electrical signals on dozens of lanceolate endings surrounding the hair follicle sum together at the spike initiation site, located at the base of the lanceolate endings, to reach a threshold needed to trigger the generation of impulses, or spikes. These impulses are then conveyed in an all-or-none manner, all the way to the spinal cord and brain. The Aβ RA-LTMRs are so sensitive to hair follicle movement that even the slightest amount of skin indentation also triggers movement of the hair follicle relative to the lanceolate endings and thus an impulse generated at the spike initiation site. In this way, Aβ RA-LTMRs fire one or a few impulses during the onset of hair deflection, but not during the sustained phase, and then again, typically, at the offset (removal) of the deflection as the follicle regains its original position, thereby stretching lanceolate endings located on the other side of the follicle to trigger an impulse. Thus the structure of an Aβ RA-LTMR's longitudinal lanceolate endings and their intimate relationship with hair follicle epithelial cells explain not only its extraordinary sensitivity to hair deflection and gentle skin indentation, but also why it stops sending impulses to the central nervous system (CNS) during sustained phases of hair bending and skin indentation.

A second touch receptor, the Aβ SAr-LTMR, is implicated in the perception of *form* of objects held in your hand or touching your body. Aβ SAr-LTMR endings are found in both hairy and nonhairy (glabrous) skin and at highest density in skin regions endowed with the greatest tactile acuity, such as our fingertips.[5] In contrast to the Aβ RA-LTMR, the Aβ SAr-LTMR, as its name implies, is slowly adapting (SA) to sustained indentation of the skin. Also unlike Aβ RA-LTMRs, Aβ SAr-LTMRs are entirely indifferent to hair deflections; one could bend hairs until the cows came home and Aβ SAr-LTMRs wouldn't generate a single impulse. Rather, Aβ SAr-LTMRs care about skin indention, and these touch receptors send impulses to the CNS, hundreds per second, as long as the skin is indented. Structural features and unique epidermal partners of Aβ SAr-LTMRs explain this unique behavior. Each Aβ SAr-LTMR sends a single process into the skin, where it branches extensively in a small region of skin, and each branch terminates in close apposition to a curiously interesting, specialized skin cell termed a Merkel cell (named after the German scientist Friedrich Sigmund Merkel, who first described them in 1875). In hairy skin, each Aβ SAr-LTMR forms close connections with 20–40 Merkel cells, which are clumped together with adjacent skin cells into a structure we call a "touch dome."[6] Merkel cell clusters found in glabrous skin are more numerous and closer together than those in hairy skin, reflecting a higher density of innervation of glabrous skin by Aβ SAI-LTMRs; this feature underlies the high tactile acuity of our fingertips. Recent discoveries have revealed a simple mechanism to explain why Aβ SAI-LTMRs are highly sensitive to skin indentation and continue to send impulses to the CNS as long as skin is indented. It was found that the Merkel cell itself is mechanically sensitive, and it conveys a signal, the nature of which is still being worked out, to the adjacent Aβ SAI-LTMR ending.[7] The immediate burst of Aβ SAI-LTMR impulses during the initial, dynamic phase of skin indentation is likely due to direct mechanical stimulation and excitation of the Aβ SAI-LTMR itself, and the second, prolonged phase of Aβ SAI-LTMR impulse generation during a sustained indentation is mediated by Merkel cell compression and indirect excitation of the Aβ SAI-LTMR. In this way, the Aβ SAI-LTMR is highly sensitive to indentation, reporting on object form, and the Merkel cell–Aβ SAI-LTMR mode of signal propagation provides a means of continuous (slowly adapting) reporting of indenting mechanical stimuli. This is what you need to read Braille dots or distinguish the ridges on the edge of a dime when you fish one out of your pocket.

Morphological oddity also distinguishes, and helps explain, a third touch receptor type, the Aβ Field-LTMR, whose endings are found exclusively in the hairy skin of mammals, including us. Aβ Field-LTMRs are highly sensitive to gentle stroking across hairy skin, yet they are curiously insensitive to innocuous skin indentation and hair deflection. The terminal morphologies of Aβ Field-LTMRs are expansive; each Aβ Field-LTMR extends a single process to the skin, where it branches profusely, giving rise to up to 200 terminals, each wrapping circumferentially, and typically two times, around a hair follicle.[8] Unlike the Aβ RA-LTMR's longitudinal lanceolate endings, the Aβ Field-LTMR's circumferential endings are located at a distance from hair follicle epithelial cells, and thus they do not form intimate contacts with the hair follicles they encircle. They also do not associate directly with the mechanically sensitive Merkel cell, thereby explaining their relative insensitivity to hair deflection and skin indentation. In fact, electrical recordings revealed that each of the 200 or so circumferential endings of an individual Aβ Field-LTMR is a distinct, relatively insensitive mechanosensory unit. Another notable feature of the Aβ Field-LTMR is that its spike initiation site is not at the extreme ending, as it is for Aβ SAI-LTMRs and Aβ RA-LTMRs. Rather, it is located at process branch points far from the individual circumferential endings that wrap loosely around hair follicles. These structural features support a model in which stroking across a large area of skin activates many weakly mechanosensitive circumferential endings to cause a large number of weak, nonimpulse generating waves of electrical excitation that summate at distantly located spike initiation sites to reach a threshold for triggering an all-or-none impulse that propagates from the skin to the spinal cord and brain. In this way, Aβ Field-LTMRs, which remain silent during deflection of individual hairs or gentle indentation of small regions of skin, transmit bursts of impulses during stroking across large swaths of skin (see figure 8).

We see that form underlies function in touch sensation. It is noteworthy that the morphologies and patterns of touch receptor endings vary across skin regions; the fingertips have a different constellation and organization of touch receptor ending types compared to the hairy skin of the back, and both of these skin regions are distinguished from the forms and patterns in penis skin or tongue skin. Indeed, this uniqueness underlies the specialized functions of different skin regions of the body. Glabrous skin on our fingertips is specialized for high acuity, enabling us to read Braille, for example. Whereas hairy skin on our backs is poised for high sensitivity, the skin on our tongues is specialized to distinguish the textural components of the foods that we may or may not want to eat. Accordingly, the morphologies of cutaneous endings of the distinct classes of the touch receptors, and their unique patterns of arrangement across the different body regions, allow us to extract diverse features of the physical world for many purposes. The brain, for its part, receives, or listens to, touch receptor ensembles emanating from the skin and, after combining experience and internal state, interprets what it hears. In this way, the exquisite ending structures of our neural instruments of touch, the touch receptors, and their unique combinations and arrangements in different skin regions, enable myriad orchestral ensembles of impulses to flow from the skin to the brain to produce a rich array of touch percepts.

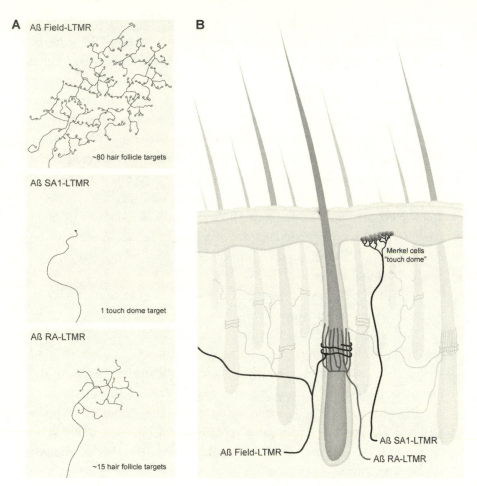

FIGURE 8. Three types of touch receptors and their endings in the skin. (A) The endings of a single Aβ Field-LTMR, an Aβ SA1-LTMR, and an Aβ RA-LTMR in a top-down view of the skin. (B) Each Aβ Field-LTMR forms circumferential endings that wrap around many hair follicles, allowing this touch receptor to respond to gentle stroking across the skin. Aβ SA1-LTMRs form endings with a single cluster of specialized skin cells, called Merkel cells, enabling it to respond as long as the skin is indented. Aβ RA-LTMRs form longitudinal endings that make intimate contacts with 10–15 hair follicles, rendering this touch receptor highly sensitive to the bending of hairs.

NOTES

1. V. E. Abraira and D. D. Ginty, "The Sensory Neurons of Touch," *Neuron* 79 (2013): 618–639.
2. A. Zimmerman, L. Bai, and D. D. Ginty, "The Gentle Touch Receptors of Mammalian Skin," *Science* 346 (2014): 950–954.
3. L. Bai et al., "Genetic Identification of an Expansive Mechanoreceptor Sensitive

to Skin Stroking," *Cell* 163 (2015): 1783–1795; L. Li et al., "The Functional Organization of Cutaneous Low-Threshold Mechanosensory Neurons," *Cell* 147 (2011): 1615–1627; M. Rutlin et al., "The Cellular and Molecular Basis of Direction Selectivity of Aδ-LTMRs, *Cell* 159 (2014): 1640–1651.

4. L. Li and D. D. Ginty, "The Structure and Organization of Lanceolate Mechanosensory Complexes at Mouse Hair Follicles," *Elife* 3 (2014): e01901.

5. K. O. Johnson and S. S. Hsiao, "Neural Mechanisms of Tactual Form and Texture Perception," *Annual Review of Neuroscience* 15 (1992): 227–250.

6. Y. S. Doucet, S. H. Woo, M. E. Ruiz, and D. M. Owens, "The Touch Dome Defines an Epidermal Niche Specialized for Mechanosensory Signaling," *Cell Reports* 3 (2013): 1759–1765.

7. S. H. Woo, E. A. Lumpkin, and A. Patapoutian, "Merkel Cells and Neurons Keep in Touch," *Trends in Cell Biology* 25 (2015): 74–81.

8. Bai et al., "Genetic Identification of an Expansive Mechanoreceptor Sensitive to Skin Stroking."

The Bane of Pain Is Plainly in the Brain

Allan Basbaum

STUB YOUR TOE and your toe hurts—a lot. So the pain is clearly in your toe. Or is it? Consider the following: every patient who loses a limb will experience a phantom limb. However, only some of them will feel intense, debilitating pain in the phantom. Where is the pain in this condition? Another interesting paradox is demonstrated with use of the thermal grill. This device consists of alternating warm and cool metal bars. Not surprisingly, if you place your hand on the grill when the warm and cool bars are activated separately, you will experience warm and cool sensations respectively. However, when the warm and cool bars are turned on together, most individuals will feel intense, burning pain. And they will reflexively quickly withdraw their hands. With the thermal grill, there is pain in the absence of "painful" stimuli; it is an illusion of pain.

What can explain these puzzling pain experiences? The answer is that pain is a complex perception that is generated in the brain, a perception that has both a sensory-discriminative component, which conveys the location and intensity of the pain, and an emotional-contextual component, which gives pain its characteristic unpleasant quality. Just as there is nothing inherently beautiful in a Mondrian painting (many, but certainly not all, people perceive it as such), so there is nothing inherently painful in the limb in which the pain is felt. That is certainly the case for phantom limb pain. The perception of pain in a phantom limb also reflects the persistence of a neural map that represents all body parts, including the missing limb, in a region of the brain called the somatosensory cortex. Whether or not pain is perceived depends on when, where, and how the electrical signals from neurons in the limb are processed by the brain and the context in which the pain occurs.

So how is "normal" pain generated when you burn your hand, break a leg, or undergo a major surgical procedure? We have considerable information about the neural circuits that transmit the injury messages from the injury site to the spinal cord and from there to the brain, where the perception of pain is generated.[1] Noxious, potentially tissue-damaging stimuli (thermal, mechanical, or chemical) activate distinct subsets of nerve fibers in the limbs, called nociceptors. These nociceptors express molecularly distinct receptors that respond specifically to different types of stimuli. For example, heat activates nerve endings that express a heat-sensitive receptor called TRPV1, and cool stimuli activate a different receptor called TRPM8. It is interesting that natural products also activate these receptors. Thus capsaicin, the painful ingredient in hot peppers, which when eaten evokes pain, selectively activates TRPV1. The cooling sensation evoked by menthol results from activation of TRPM8. Painful mechanical stimuli, like pinches, target yet other types of nociceptor nerve endings in the skin.

Axons, the information-sending fibers of nociceptive neurons, travel to the spinal cord, where they enter circuits that transmit their message to higher brain structures. Some spinal-cord-to-brain pathways carry sensory-discriminative information about the stimulus (location and intensity). Other pathways engage different regions of the brain that process and generate emotions. Eventually the information reaches the cerebral cortex, where the percept of pain, which results from integration of the sensory-discriminative and emotional aspects, is generated. Often forgotten, but also important, are the cognitive, experiential features that influence pain perception. In other words, the situation in which the noxious stimulus occurs can profoundly influence the ultimate perception. Think for a moment about the athlete who is injured in the middle of an intense match but experiences the pain only when the match is over.

As there have been great strides in brain imaging technology, one might assume that identifying the cortical areas where pain is generated should be pretty straightforward. Place a participant in a scanner, image the brain using functional magnetic resonance imaging (fMRI), and map the activated areas of the brain in response to a painful stimulus. Unfortunately, here is where things get complicated. Yes, we can identify areas of activity in the somatosensory cortex that correlate with the intensity and location of the stimulus. And in many other brain areas, such as the anterior cingulate gyrus and the anterior insular cortex, activity correlates with the emotional components of pain—that is, the unpleasantness of the experience. But we still cannot pinpoint the brain areas the activity of which unequivocally signals the quality, intensity, and emotional impact of a subject's pain.

These problems are not merely of academic interest. The question of pain localization in the brain is certainly relevant to acute pain—namely, pain provoked by an injury stimulus—or to short-term postoperative pain. But the ability to image pain objectively is most relevant when it comes to chronic pain, which is defined as pain that lasts more than three months, as occurs with arthritis, painful diabetic neuropathy, postherpetic neuralgia, back pain, or cancer pain. As pain is a subjective experience, the physician must rely solely on the patient's report. Determining how best to treat the pain, including how much analgesic medication to administer, would be greatly facilitated if there were an objective measure of a patient's pain.

Let's go back to the thermal grill—those alternating warm and cool metal bars. If you image the brain of an individual who experiences the illusion of pain created by the thermal grill, surprisingly you observe a pattern of brain activity in regions normally associated with both the sensory-discriminative and emotional aspects of pain. In fact, the pattern is very similar to the activity pattern observed in someone who places a hand on a hot object.[2] Remember that the thermal grill activates only nerve fibers that respond to harmless warm and cool stimuli. The nociceptors (the pain fibers) are not activated. Apparently, you do not need activity of pain fibers in the hand to produce pain or to evoke brain activity normally generated by activation of these fibers. Rather, it is the activity of the brain that generates the subjective magnitude, location, and unpleasantness of the pain. Should we conclude that the pain is not really in the hand, even though the participant certainly experiences pain there? In a similar vein, phantom limbs obviously persist even when the limb no longer exists. In this case, it is thought that preexisting maps of the limb in the cerebral cortex create the illusion. So when an individual reports pain in a phantom limb, the pain must clearly be somewhere else. Where the pain is and how it arises remain a mystery.

Consider yet another interesting paradox. Nerve fibers communicate with others by conduction of electrical signals along their axons, resulting in the release of neurotransmitters that transmit information from one neuron to the next. Conduction of these electrical signals involves activation of ion channels, notably a large class of channels that pass sodium ions into the neuron, resulting in propagation of the electrical signal. When a dentist administers a local anesthetic, the anesthetic blocks all sodium channels, thus preventing nerve fiber conduction of signals from the oral cavity and teeth to the brain. As a result, pain perception is eliminated. In fact, all sensations are eliminated because local anesthetics nonselectively block conduction in all types of nerve fibers, including those that respond to touch and joint movement, not only the nociceptors.

The ideal local anesthetic, of course, would block only the sodium ion channels relevant to nociceptors. Enter the Nav1.7 subtype of sodium channel. Not too long ago, a genetic analysis of several members of Pakistani families who never experienced pain revealed that these individuals had a mutation of the gene that codes for the Nav1.7 channel, resulting in a channel that is inoperable.[3] These individuals were cases of congenital insensitivity to pain. Subsequent studies demonstrated that the Nav1.7 subtype is highly enriched in nociceptors, a finding that suggested that the channel may be uniquely associated with pain generation. Not surprisingly, many pharmaceutical companies are working to develop drugs that selectively target the Nav1.7 subtype, as there could be preferential relief of pain, with limited adverse side effects associated with many other pain-relieving drugs, such as opioids.

The Nav1.7 subtype of channels is especially relevant to the question posed in this essay. A recent experiment compared the pattern of brain imaging in response to an intense mechanical (normally pain-evoking) stimulus in two patients with congenital insensitivity to pain and in four healthy individuals. The result was surprising. The pattern of brain activity in the two pain-free subjects did not differ from that evoked by the same painful stimulus in the controls.[4] In other words, a comparable pattern of brain activity can be generated when pain is experienced in response to a frankly painful stimulus; when there is an illusion of pain produced by a nonpainful stimulus (with the thermal grill); and even in response to a normally painful stimulus in an individual who never experiences pain. Apparently, induction of brain activity in what some researchers call the "pain matrix," which includes areas implicated in the sensory-discriminative and emotional features of the pain experience, is not sufficient to generate the experience of pain.

Consider the following hypothetical experiment, in which an individual with a loss of function mutation of Na$v1.7$ places a hand on the thermal grill. Remember that the sodium channels in the peripheral nerve fibers that are activated by warm and cool stimuli function normally in these individuals, as do the nerve fibers that carry the warm and cool signals to the spinal cord. When only the warm bars are engaged, we expect that a warm sensation will be reported. Turn off the warm bars and engage the cool bars, and this individual should experience cooling. But what do you predict will happen when the interlaced warm and cool bars are turned on at the same time?

Assuming that there are no alterations in the brain of the individuals with congenital insensitivity to pain, then the brain should be able to process the information and generate the patterns of neural activity that underlie the illusion of pain. But will this activity pattern in the brain resemble that generated for real pain? An even more interesting question, perhaps, is whether these individuals will for the first time experience pain, or at least the illusion of pain. Of course, they may not use the same words to describe the experience, but would there be an emotional dimension to their experience, which is among the most important features of the pain experience? Will their heart rates increase? Will they pull their hands away quickly, as do most individuals who place their hand on the thermal grill?

Despite this apparent disconnect between patterns of brain activity and the experience of pain, recent studies suggest that the use of brain imaging to monitor pain should not be abandoned. Using functional magnetic resonance imaging, Tor Wager and colleagues reported that they have identified an extensive pattern of brain activity in individuals that could be used to predict the magnitude of acute pain that they experienced in response to an intense heat stimulus.[5] Not only did this "signature" of acute pain—which bears much resemblance to the "pain matrix"—reliably differ from that provoked by a warm stimulus, but the magnitude of the signature was also reduced by an opioid, remifentanil, which reduces acute pain. These findings provide an important step forward, but as noted above, developing an objective measure of chronic pain is where the greatest need lies. To this end, Vania Apkarian and colleagues monitored brain neural activation patterns in thirty subacute back pain patients (those with pain that has been present for at least six weeks but less than three months) who either recovered from or transitioned to chronic back pain.[6] After tracking these patients for three years, the authors concluded that preexisting anatomical and functional network connections among specific brain areas—notably the medial prefrontal cortex, the amygdala, and the nucleus accumbens—reliably predicted whether an individual patient's pain did or did not resolve beyond the subacute state. The authors posed the intriguing hypothesis that relief from a subacute back pain condition is a rewarding experience that may not be inducible in patients in whom these connections are abnormal. Although these studies did not identify the brain area where pain is experienced, they highlight the complexity of the brain circuitry that underlies the development of chronic pain.[7] These findings also suggest that it is not a single brain region, but perhaps a group of brain regions that no longer share information as they once did, that mediate chronic pain.

So where are we when it comes to answering the question, "Where is the pain?" The pain percept is unquestionably processed in your brain. Indeed, I continue to tell my students that the Bane of Pain Is Plainly in the Brain. But we still don't know where. It is interesting that neurosurgeons have for years tried to ablate discrete brain regions so as to eliminate chronic pain, unfortunately with little success. Furthermore, although electrical stimulation of the brain in patients undergoing ablative brain surgery for uncontrollable seizures can evoke a variety of perceptions (colors, sounds, tingling, even memories), it is very difficult to evoke pain.[8] Until we identify a discrete brain area(s) that encodes the experience of pain, we will continue to depend on the patient's verbal report. For now, I guess, the pain is in the toe.

NOTES

1. A. I. Basbaum, D. M. Bautista, G. Scherrer, and D. Julius, "Cellular and Molecular Mechanisms of Pain," *Cell* 139 (2009): 267–284.

2. A. D. Craig, E. M. Reiman, A. Evans, and M. C. Bushnell, "Functional Imaging of an Illusion of Pain," *Nature* 384 (1996): 258–260.

3. J. J. Cox, F. Reimann, A. K. Nicholas, et al., "An SCN9A Channelopathy Causes Congenital Inability to Experience Pain," *Nature* 444 (2006): 894–898.

4. T. V. Salomons, G. D. Iannetti, M. Liang, and J. N. Wood, "The 'Pain Matrix' in Pain-Free Individuals," *Journal of the American Medical Association: Neurology* 73 (2016): 755–756.

5. T. D. Wager, L. Y. Atlas, M. A. Lindquist, et al., "An fMRI-Based Neurologic Signature of Physical Pain," *New England Journal of Medicine* 368 (2013): 1388–1397.

6. E. Vachon-Presseau, P. Tetreault, B. Petre, et al., "Corticolimbic Anatomical Characteristics Predetermine Risk for Chronic Pain," *Brain* 139 (2016): 1958–1970.

7. M. A. Farmer, M. N. Baliki, and A. V. Apkarian, "A Dynamic Network Perspective of Chronic Pain," *Neuroscience Letters* 520 (2012): 197–203.

8. L. Mazzola, J. Isnard, R. Peyron, and F. Mauguiere, "Stimulation of the Human Cortex and the Experience of Pain: Wilder Penfield's Observations Revisited," *Brain* 135 (2012): 631–640.

Time's Weird in the Brain—That's a Good Thing, and Here's Why

Marshall G. Hussain Shuler and Vijay M. K. Namboodiri

TIME IS WEIRD in the brain. It flies when you're having fun and crawls when you are waiting in line at the Department of Motor Vehicles. So, while it's natural to assume that our perception of time should accurately reflect reality, our experience in everyday life does not bear this out. Time also often sways our decisions inordinately; given the option between $100 to be received in a year or $110 in a year and a week, people often pick the latter, but after a year passes, most would rather have the $100 immediately than wait a week for the additional $10. Thus our ability to comprehend, remember, generate, and factor time into decisions exposes that our mental world seemingly cares little about its faithful adherence. Why should that be? Surely evolutionary success would benefit from representing time *as it is*. Or would it? Until now, attempts to explain our distorted sense of time have reasoned that its quirky nature is but nature's flawed attempt to represent time faithfully. But what if it takes the form it does in the service of a greater objective? Perhaps our timing sense and its ostensible failings can be recast not as bugs but rather as features permitting appropriate decision making.

The subjective perception of time has been the subject of careful scientific investigation.[1] Indeed, our qualitative sense that something is amiss with our perception of time has been rigorously measured using diverse means to address the many ways in which we, mere humans (and our animal friends), misrepresent time. We have come to understand that our sense of time is not just randomly inaccurate, but rather that it fails in a particular manner that exhibits systematic deviations from objective time. For instance, when asked to estimate the midpoint of a time interval, we typically place the midpoint too early.[2] Relatedly, when we are asked to estimate the duration of an interval, the variability in our responses is often observed to increase in direct proportion to the interval to be timed.[3] These errors in accuracy (how close our average estimate is to reality) and precision (the variability about that estimate), respectively, are key characteristics of our timing sense. Therefore, attempts to understand our subjective timing sense must aim to explain these phenomena.[4] However, it remains unclear *why* our subjective sense of time should take a form that results in these particular systematic errors. What purpose does it serve?

Before addressing this central question, we must appreciate that time is not given to us from our environment; rather, our brains must create it. Though we become cognizant of events in the world through our senses, without a means to remember and relate events that are disparate in time, our experience of the world would be ephemeral. But it's not *just* about being able to remember the sequence of experienced events. An appreciation of time demands that the *intervals* between events are encoded in the brain. It is by dint of our brain's ability to manufacture time—to represent and remember those created intervals and to reproduce those intervals—that we have any appreciation of time at all. Since it's up to the brain to create time, what's wrong with the seemingly rational notion that the brain should represent time as it is—that is, with little error, being both accurate and precise?

To begin to understand why time might be created in the brain in any other way, let us ponder why it's beneficial to appreciate time in the first place: to inform our decisions.[5] Since our success as a species depends on how well we are able to choose our actions based on their expected outcomes (rewards or punishments), it seems reasonable that our brains would have evolved effective means of learning the relationships between sensory events (that is, a sound, light, or odor) and the outcomes they predict. For instance, a visual cue (sighting of prey) may foretell a reward—if the reward is pursued—of a given magnitude (prey size) after a certain delay (travel and capture time). Having learned that a particular cue means reward, the predator can then evaluate the worth of that opportunity by weighing its benefit against the cost of time needed for its attainment. Such decisions that weigh a reward by its cost in time are also commonplace for humans (for example, should I go to a nearby restaurant that's okay or a great but more distant restaurant?). But what *is* the cost of time?

Numerous fields have labored to address what the cost of time *ought* to be—that is, how time should be weighed for a stated goal so that the goal is achieved.[6] A goal from the field of ecology that has great appeal is that animals forage while in an environment so as to optimize the amount of reward obtained. An algorithm that has been proposed to achieve this goal asserts that the worth of an offering of delayed reward *(r)* is determined by normalizing by the time *(t)* needed in its attainment—that is, the offer's rate of reward, *(r/t)*.[7] But were you offered an opportunity to wait in a short line to receive a small reward or a long line to receive a large reward, would you always choose the short line if it yielded the highest rate of reward? Now, curiously, that depends! It depends on the reward rate one may expect to receive when *not* standing in line. Let's call this the background reward rate and examine why it should influence our decision as to which reward option to choose.

Consider the difference in time needed to obtain the large-later versus the small-sooner reward. Over that amount of time, how much reward would you have received given the background reward rate? This amount is the *opportunity cost* of time of standing in the large-later rather than the small-sooner line (the expected loss of background reward given the additional time spent standing in the long line). The question, then, is—for the time spent in the long line to receive its reward—is it better to invest time in choosing the short line to collect its reward plus that amount expected for *not* standing in line the remaining time? When the background reward rate is low (background rate is less than difference in the offered rates), choosing the larger-later line yields more reward, but when the rate of reward of not standing in line is high (background rate is greater than difference in the offered rates), choosing the smaller-sooner line yields more reward. This approach of evaluating opportunities from the perspective of the opportunity cost of time similarly affords a means of determining if it's even worth entering a line for reward at all.

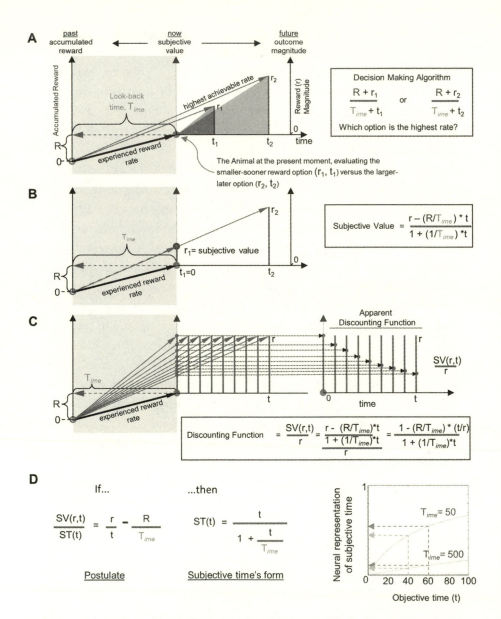

A

past
accumulated
reward ← now
subjective
value → future
outcome
magnitude

Look-back
time, T_{ime}

highest achievable rate

r_2

r_1

Accumulated Reward

R

0-

experienced reward
rate

t_1 t_2 time

Reward (r)
Magnitude

0

Decision Making Algorithm

$$\frac{R + r_1}{T_{ime} + t_1} \quad or \quad \frac{R + r_2}{T_{ime} + t_2}$$

Which option is the highest rate?

The Animal at the present moment, evaluating the
smaller-sooner reward option (r_1, t_1) versus the larger-
later option (r_2, t_2)

B

T_{ime}

r_2

r_1= subjective value

0

R

0-

experienced reward
rate

$t_1=0$ t_2

$$Subjective\ Value = \frac{r - (R/T_{ime}) * t}{1 + (1/T_{ime}) * t}$$

C

T_{ime}

r

R

0-

experienced reward
rate

t

Apparent
Discounting Function

r

$$\frac{SV(r,t)}{r}$$

0 time t

Discounting Function $= \dfrac{SV(r,t)}{r} = \dfrac{\frac{r - (R/T_{ime})*t}{1 + (1/T_{ime})*t}}{r} = \dfrac{1 - (R/T_{ime}) * (t/r)}{1 + (1/T_{ime})*t}$

D

If...

$$\frac{SV(r,t)}{ST(t)} = \frac{r}{t} - \frac{R}{T_{ime}}$$

Postulate

...then

$$ST(t) = \frac{t}{1 + \dfrac{t}{T_{ime}}}$$

Subjective time's form

Neural representation
of subjective time

1

$T_{ime}= 50$

$T_{ime}= 500$

0 20 40 60 80 100

Objective time (t)

FIGURE 9. Schematic of the TIMERR algorithm. (A) The TIMERR decision-making algorithm
(right) asserts that the worth of reward offers is evaluated by adding the reward rate of an offer
(defined by the magnitude of reward, r, and its associated delay, t) to the experienced reward
rate of the environment (defined by the total accumulated reward, R, collected over the time
the agent "looks back" into its past, T_{ime}) and selecting that which is greatest. This algorithm
can be depicted graphically (left), as an animal (filled gray circle) evaluating which offered
reward yields the highest achievable reward rate from the perspective of looking back at an
interval of time (T_{ime}) over which R rewards have been accumulated. (B) The subjective value

Were the background reward rate greater than the reward rate of an offer, one should forgo that opportunity of reward, as more is to be had in lieu of taking that offer. Therefore, obtaining the greatest reward while foraging in an environment requires evaluating the worth of an offer with respect to its opportunity cost of time. But how should the opportunity cost of time be determined?

Recently, we have proposed a decision-making algorithm called TIMERR (Training Integrated Maximized Estimate of Reward Rate), which seeks to maximize the amount of reward received by an animal while foraging in an environment.[8] A key proposition of this algorithm is that an animal looks back an interval of time into its recent past to determine the rate of reward of the environment. In this way, the animal evaluates an offered reward by assessing whether, if the reward is pursued, the resulting reward rate of the environment will be greater than that which has already been achieved (figure 9A). This simple algorithm can be rewritten to form an expression for how much a reward of a given magnitude and delay should be worth—its so-called subjective value (figure 9B). A delayed reward's subjective value is the amount of immediate reward that is treated as being equally good. When TIMERR is viewed in this manner, it's evident that an opportunity cost of time needed to

of a delayed reward (being the magnitude r1 of an immediate reward t = 0, which is treated as equal to that of a delayed reward r2, t2), can be derived from the TIMERR algorithm (right) and visualized graphically (left) as the y-axis intercept of the delayed reward when determining its resulting reward rate. It is important that the opportunity cost of time (R/T_{ime} * t) is taken into account as the experienced rate of reward (R/T_{ime}) times the time invested to obtain the reward offer (t). (C) How the subjective value of a reward of a given magnitude diminishes as the time invested to obtain it increases can be re-expressed as a discounting function of time (inset) and graphically depicted by replotting the subjective values of a reward displaced at equal intervals into the future (left, black circles) at their respected delays (right). The decrement in subjective values forms a hyperbolic function as often observed experimentally. (D) Postulating that the subjective reward rate is equal to the difference between the offered reward rate (expected reward divided by the expected delay) and the experienced reward rate (left), the TIMERR algorithm can be used to derive an expression (middle) for the subjective representation of time, ST(t). Normalizing ST(t) by T_{ime} yields a bounded, concave function of objective time (right), whose curvature becomes more pronounced the shorter the look-back time (T_{ime} = 50) and more linear to objective time the larger the look-back time (T_{ime} = 500).

obtain a given reward is being subtracted from the reward's magnitude. More generally, how the subjective value of any delayed reward decreases with time can be expressed as a temporal discounting function. When expressed as a temporal discounting function, the subjective value of a reward diminishes in TIMERR in a manner that is dependent on the amount of time the agent looks back to estimate the environment's reward rate (figure 9C). Since the worth of an offer is evaluated from the perspective of how much time the animal looks back into its past, the consequence is that the greater the look-back time, the more patient the animal appears to be (being willing to wait a longer time for the same reward).

Although this algorithm formalizes how an animal ought to behave in its effort to maximize reward, how well does it account for established decision-making behaviors that factor time? TIMERR, in fact, reconciles a long list of seemingly disparate yet commonly observed decision-making behaviors, many of which are interpreted as evidence that humans and animals are not rational decision makers.[9] Therefore, TIMERR recasts these ostensibly irrational behaviors as the rather rational decision making of an animal operating under the constraint of what is known (and knowable) of an uncertain environment.

Does this explain why time should be represented in any way other than objectively in the brain? Under TIMERR, we have reasoned that the subjective reward rate of an offer—defined as the ratio of its subjective value to the subjective representation of the delay—is equal to the amount that the offer's reward rate is higher than the experienced reward rate of the environment (figure 9D, left panel). This is to say that the subjective reward rate accurately depicts the change in the objective reward rate. Looked at another way, this postulate is equivalent to stating that the reduction in subjective value due to a delay is linearly proportional to the subjective representation of that delay. From this, an expression for the subjective representation of time can be derived from the TIMERR algorithm (figure 9D, middle panel). It is striking that the resulting representation of subjective time is a concave function of objective time—that is, as objective time increases, subjective time increases at a progressively slower rate. If the representation of subjective time is by a limited neural resource (as is commonly reasoned), this concave function is bounded (figure 9D, right panel) so that no utilization of this neural resource corresponds to zero delay, and full utilization corresponds to an indefinitely long interval. It is important that the amount of nonlinearity (curvature) in this function is controlled by the look-back time: the longer the look-back time, the closer the representation is to objective time, whereas the shorter the look-back time, the more subjective time bends away from objective time (figure 9D, right panel). Therefore, according to TIMERR, the neural representation of subjective time takes the form that it does—deviating from objective time—in the service of making decisions that increase the rate of reward received in a world of uncertain future reward.

While TIMERR provides a rationale for why the representation of time should take a particular form and why that needn't be linearly related to objective time, how does it explain timing behaviors that reveal systematic errors in accuracy and precision that are commonly observed in humans and animals? With respect to accuracy, when asked to estimate the midpoint between two intervals, subjects often report this midway point as being earlier than the mean of the two intervals. This is curious in that if subjective time was accurate—even if imprecise—one would expect the typically reported midpoint to be the *arithmetic* mean between the two intervals. It is interesting that the midpoint can actually be defined mathematically in two ways other than the arithmetic mean, and, perhaps, were reports to accord with one or another definition, some insight into how the brain represents time might be gained. The *geometric mean* is the square root of the product of the intervals and is lower than or equal to the arithmetic mean, while the *harmonic mean* is the *reciprocal of the mean of reciprocals of the intervals* and is lower than or equal to the geometric mean. While subjects most commonly report a midpoint approximating the geometric mean,[10] reports actually span a range from the harmonic to the arithmetic mean, making them difficult to reconcile within most prior theories of timing.[11] TIMERR, however, predicts that the reported midpoint would in fact range from the harmonic to the arithmetic mean, depending on how much time the animal looks back into its past to estimate reward rate. Not looking back at all results in bisecting an interval as its harmonic mean, while increasing the look-back time leads to bisection approaching the geometric mean, after which further increases lead to bisections that approach the arithmetic mean. The amount of time used to estimate the experienced reward rate of the environment thus controls the degree of curvature of the (subjective) neural representation of time, which in turn affects the report of the midway point. In this way, TIMERR reconciles the diversity of reports of the midway point while explaining why such inaccuracies in timing are observed at all.

But why do errors in timing grow in proportion to the interval to be timed? One way to build a clock is to create a timing mechanism that ticks with perfect regularity. Accumulating a count of these ticks would yield temporal intervals of absolute accuracy and precision. However, the brain possesses only noisy information processors in neurons, incapable of producing perfectly rhythmic pulses. How, then, can noisy neural processors produce temporal intervals?[12] To represent time subjectively, the way TIMERR postulates,[13] it turns out that the neural processors need to have feedback onto themselves.[14] A side effect of this feedback is that it causes an amplification of the noise in processing, such that the longer the duration, the larger the resultant noise. Thus representing time subjectively using the TIMERR postulate automatically results in errors that scale with the interval being timed, as widely observed. It is interesting that because TIMERR postulates that the representation of time is tied to the look-back time used to calculate background reward rate, it predicts that changes in the look-back time will affect our perception of time. Specifically, when the reward rate is high (you are having fun), it is predicted that the look-back time will shorten and the perception of time will be shorter (time flies). In other words, time flies when you're having fun.

The model proposed here has the potential to rationalize our warped sense of time as a natural consequence of reasonable decision making that accounts for the costs associated with time. By doing so, this theory is capable of explaining many established findings in time perception (while providing new, falsifiable predictions yet to be experimentally tested). Thus the idiosyncratic nature of our sense of time is not without its merits after all. And that, apparently, is why time is weird.

NOTES

1. M. Bateson and A. Kacelnik, "Rate Currencies and the Foraging Starling: The Fallacy of the Averages Revisited," *Behavioral Ecology* 7 (1996): 341–352; J. Gibbon, "Scalar Expectancy-Theory and Weber's Law in Animal Timing," *Psychological Review* 84 (1977): 279–325; P. R. Killeen and J. G. Fetterman,

"A Behavioral Theory of Timing," *Psychological Review* 95 (1988): 274–295; M. S. Matell and W. H. Meck, "Neuropsychological Mechanisms of Interval Timing Behavior," *Bioessays* 22 (2000): 94–103.

2. G. L. Chadderdon and O. Sporns, "A Large-Scale Neurocomputational Model of Task-Oriented Behavior Selection and Working Memory in Prefrontal Cortex," *Journal of Cognitive Neuroscience* 18 (2006): 242–257; R. M. Church and M. Z. Deluty, "Bisection of Temporal Intervals," *Journal of Experimental Psychology Animal Behavior Processes* 3 (1977): 216–228; C. D. Kopec and C. D. Brody, "Human Performance on the Temporal Bisection Task," *Brain and Cognition* 74 (2010): 262–272; J. M. Levy, V. M. Namboodiri, and M. G. Hussain Shuler, "Memory Bias in the Temporal Bisection Point," *Frontiers in Integrative Neuroscience* 9 (2015): 44.

3. J. Gibbon, R. M. Church, and W. H. Meck, "Scalar Timing in Memory," *Annals of the New York Academy of Sciences* 423 (1984): 52–77; J. H. Wearden and H. Lejeune, "Scalar Properties in Human Timing: Conformity and Violations," *Quarterly Journal of Experimental Psychology* (Hove) 61 (2008): 569–587.

4. M. Bateson, "Interval Timing and Optimal Foraging," in *Functional and Neural Mechanisms of Interval Timing*, ed. W. H. Meck (Boca Raton: CRC Press, 2003), pp. 113–141; X. Cui, "Hyperbolic Discounting Emerges from the Scalar Property of Interval Timing," *Frontiers in Integrative Neuroscience* 5 (2011): 24; A. Kacelnik and M. Bateson, "Risky Theories—The Effects of Variance on Foraging Decisions," *Integrative and Comparative Biology* 36 (1996): 402–434; D. W. Stephens, "Discrimination, Discounting and Impulsivity: A Role for an Informational Constraint," *Philosophical Transactions of the Royal Society of London. Series B* 357 (2002): 1527–1537; G. Zauberman et al., "Discounting Time and Time Discounting: Subjective Time Perception and Intertemporal Preferences," *Journal of Marketing Research* 46 (2009): 543–556.

5. V. M. K. Namboodiri et al., "A General Theory of Intertemporal Decision-Making and the Perception of Time," *Frontiers in Behavioral Neuroscience* 8 (2014): 61.

6. G. H. Pyke, "Optimal Foraging Theory: A Critical Review," *Annual Review of Ecology and Systematics* 15 (1984): 523–575; P. A. Samuelson, "A Note on Measurement of Utility," *Review of Economic Studies* 4 (1937): 155–161.

7. D. W. Stephens and J. R. Krebs, *Foraging Theory* (Princeton, NJ: Princeton University Press, 1986).

8. Namboodiri et al., "A General Theory of Intertemporal Decision-Making and the Perception of Time."

9. Ibid.; V. Namboodiri, S. Mihalas, and M. Hussain Shuler, "Rationalizing Decision-Making: Understanding the Cost and Perception of Time," *Timing and Time Perception Reviews* 1 (2014): 4; V. Namboodiri, S. Mihalas, and M. Hussain Shuler, "A Temporal Basis for the Origin of Weber's Law in Value Perception," *Frontiers in Integrative Neuroscience* 8 (2014): 79.

10. Church and Deluty, "Bisection of Temporal Intervals"; J. Gibbon, "The Structure of Subjective Time: How Time Flies," *Psychology of Learning and Motivation* 20 (1986): 105–135.

11. P. R. Killeen, J. G. Fetterman, and L. Bizo, "Time's Causes," in *Advances in Psychology: Time and Behavior: Psychological and Neurobehavioral Analyses,* ed. C. M. Bradshaw and E. Szabadi (Amsterdam: Elsevier, 1997), pp. 79–131.

12. P. Simen et al., "A Model of Interval Timing by Neural Integration," *Journal of Neuroscience* 31 (2011): 9238–9253.

13. V. M. Namboodiri, S. Mihalas, and M. G. Hussain Shuler, "Analytical Calculation of Errors in Time and Value Perception Due to a Subjective Time Accumulator: A Mechanistic Model and the Generation of Weber's Law," *Neural Computation* 28 (2015): 89–117.

14. Neuronal firing patterns can be assumed to be a random clock (Poisson process) that ticks at a given frequency but whose pulses are independent of one another (i.e., with no periodicity). Within this assumption of noisy firing patterns, an integrator of these pulses that, in simple terms, counts the number of pulses until a target time is reached produces errors in timing that scale sublinearly with the interval being timed. This is inconsistent with observed timing behavior. It is curious that a modification of such a system that counts pulses in accordance with subjective time as represented by TIMERR requires negative feedback from the accumulator onto the clock, thereby producing errors that scale linearly with the intervals being timed. Thus accumulators abiding by TIMERR's subjective time representation naturally produce this most noticeable idiosyncrasy of time perception.

Electrical Signals in the Brain Are Strangely Comprehensible

David Foster

THIS IS THE STORY of how scientists have been using electrodes to record the activity of neurons in the brain, to measure how they respond not only to sights, smells, or sounds, but also to memories, emotions, and feelings of pleasure and pain. Neurons are special; their modus operandi is to communicate with tens of thousands of their fellow neurons, forming complex networks responsible for the intricate computations underlying all our behavior. This story spans a century of astonishing discoveries, successive waves of optimism and pessimism, and a remarkable emerging theme: the electrical signals that neurons produce, the software of our mental life, can indeed be understood.

The earliest discoveries revealed that every neuron transmits an electrical impulse, or spike, to those neurons to which it is connected. This transmission is considered an all-or-nothing event. There is no such thing as a loud spike or a quiet spike; a spike is a spike is a spike.[1] The brain makes use of these impulses to send a kind of code in which objects and events in the world are depicted in a strikingly intuitive manner. For example, the degree of stimulation of a peripheral nerve sensing the pressure on a finger is conveyed by the number of spikes recorded from that nerve in a given time period. Later was the discovery that there were neurons in the retina of the frog that spiked when and only when a small, dark disc appeared in a particular part of the visual field; activation of these "bug detector" neurons triggered a predatory behavioral response by the frog toward the position of the disc.[2] In the 1950s and 1960s, the golden age of recording from single neurons, future Nobel Prize winners David Hubel and Torsten Wiesel discovered that neurons in the visual cortex respond to features in the visual scene like sharp edges.[3] A steady progression of results elaborated the tuning of neurons in ever-deeper brain areas to more complex features, culminating in the discovery of neurons that respond selectively to hands and faces.

The triumphant mood of this era was captured in a paper by the physiologist Horace Barlow, announcing a single-neuron doctrine for perception.[4] Barlow declared that the activity of a single neuron *was* the perception, and all of mental life might be understood in terms of such activities. However, the pace of discovery of neurons tuned for specific visual features dwindled, and the approach was ripe for a challenge. David Marr was a brilliant pioneer in making mathematical models of how networks of neurons might work. (He tragically died at the age of thirty-five from leukemia.) Marr wrote a hugely influential book, *Vision,* published posthumously in 1982, and in a famous introductory chapter, he categorically questioned the whole enterprise of relating neural activity to mental processes. Here is Marr considering a neuron so selective as to respond only to the presence of one's grandmother:

> Suppose, for example, that one actually found the apocryphal grandmother cell. Would that really tell us anything much at all? It would tell us that it existed—[Charlie] Gross's hand-detectors tell us almost that—but not why or even how such a thing may be constructed from the outputs of previously discovered cells. . . . If we really knew the answers, for example, we should be able to program them on a computer. But finding a hand-detector certainly did not allow us to program one.[5]

Instead of a purely empirical approach, Marr demanded a new kind of effort: figuring out how a visual system ought to be designed, from an engineering point of view. This meant incorporating insights from the field of artificial intelligence (AI). With these insights, effective computer programs for vision could be found, and then—and only then—the details of neural responses might be interpreted.

It is ironic that just as Marr was making his call for an awareness of AI in neuroscience, AI was running out of steam. The grand idea of writing programs to achieve goals such as extracting speech from sound or objects from images was just too difficult. These are tasks that we do effortlessly but that are very difficult to explain in terms of steps in a program. Marr wanted neuroscience to pause and wait for AI to come up with the solutions, but what actually happened to AI next was surprising. A different kind of computation was developed, based on networks of artificial neurons that weren't programmed to do things explicitly but instead learned gradually by being trained on examples. This field suffered its own waves of optimism and pessimism, but now in the guise of deep learning it has reached near-human levels of performance on previously unassailable tasks like face recognition, with revolutionary applications like self-driving cars.[6]

Meanwhile, neurophysiologists continued to make important discoveries of neuronal responses, particularly in areas of the brain beyond those associated with early sensory processing. Deep in the medial temporal lobe, the neurons of the hippocampus have a particular role in memory,[7] while receiving information about the world that has passed through many different brain areas since entering in at the eye and other sense organs; likewise, the output of the hippocampus passes through many different brain areas before having any impact on actions. The hippocampus is therefore the mysterious interior of the brain; one would not expect to observe meaningful or intuitive correlates of experience in these neurons. Yet through a combination of experimental innovation and philosophical verve, this is exactly what John O'Keefe and colleagues described in the course of work beginning in the early 1970s. Allowing freely moving rats to explore space, O'Keefe reported that hippocampal neurons fired spikes only in spatially localized regions, different for each neuron. Drawing on the philosopher Kant's notion of a priori knowledge, O'Keefe and psychologist Lynn Nadel hypothesized that the neurons provided the rest of the brain with a cognitive map of space, inspiring a generation of researchers (both those who agreed and those who did not).[8]

Within this small field of hippocampal neurophysiologists, a debate raged about how spatial the hippocampal place cell responses really were. They did not seem regular enough to support the geometrical representations that some researchers wished to ascribe to them. Eventually, in several areas connected with the hippocampus, neurons were found to have exactly the kinds of geometrical responses that had been missing. First, neurons were found that responded like an internal compass to the direction an animal was facing. Then, in 2005, Edvard and May-Britt Moser and colleagues reported that cells in the entorhinal cortex, one synapse upstream of the hippocampus, respond in a hexagonal grid pattern in space—that is, according to a distinct pattern that is spatially periodic in two dimensions. These "grid cell" responses are strikingly unrelated to the behavioral trajectories of the animals, rather reflecting an internally organized structure imposed on experienced space, sometimes likened to graph paper. O'Keefe and the Mosers received the Nobel Prize in 2014 for these discoveries. Many more spatial response types have since been identified in these and related brain areas.[9]

Despite the astonishing success of single-cell neurophysiology in uncovering neural representations of the external world, there seems to be something still missing. Are our thoughts and perceptions merely responses to stimuli like the responses of single neurons? One of Marr's criticisms that remains particularly pertinent is that knowing the representation provided by a neuron is different from knowing the process at work. For example, we may know the representation provided by hippocampal place cells—they provide a map of the world—but what is the process for using a map? When you read a map, you move your eyes behind and ahead of your current location to explore where you have been and where you are going. Just knowing about place cells doesn't tell us anything about any process like this. Surprisingly, technical developments in neurophysiology may have given us an answer. After all, Barlow's single-neuron doctrine rather made a virtue of necessity in focusing on the independent activity of single neurons since experimental limitations meant that this was the only kind of data that had ever been collected. From around the 1990s, simultaneous recordings from large numbers of neurons became possible. A story that has not perhaps been told before is that the imagination of scientists ran in a somewhat different direction from where ultimately the data would take them.

Many scientists anticipated the data that would come from parallel recording methods in terms of how populations of neurons would encode the stimulus in a richer way than a single neuron acting alone. For example, it was suggested that visual neurons representing the edges of one object could fire in a synchronized way that differentiated them from the neurons representing the edges of a different object.[10] Another suggestion was that a population of neurons could represent a whole distribution of the values of some quantity rather than just one estimate.[11] What was not anticipated was that by coordinating their activity, neurons could *escape from the stimulus altogether.* Hippocampal place cells showed the way: those that fired together during behavior because they shared the same place fields came to fire together afterward during sleep. But it turned out that this was just the beginning of the story for these cells. By recording sufficient numbers of cells at the same time, scientists found that hippocampal place cells were active in sequences that corresponded to behavioral paths through the environment—literally sequences of places, but speeded up. These sequences happened in sleep, but—and this was important—they also happened while animals paused briefly in the middle of performing spatial tasks. It was as if the hippocampal map suddenly came alive: the location spotlight was wandering—mentally, as it were, without physical movement—over to other parts of space, exploring behavioral trajectories, just like our eyes moving across a map.

We now know that these sequences reflect learned information about the spatial structure of the environment. It is fascinating that the sequences can stitch together pieces of behavior that were never experienced together, creating what look like mental shortcuts. We now also know that the sequences don't just occur at random but can be directed to the current location of a goal, at the very moment an animal is about to make a movement, as if reflecting an active process of considering the path that is about to be taken.[12] Even the classical activity recorded during movement is organized into short "look-ahead" sequences, a process that becomes clear when place cells are examined in sufficiently large numbers simultaneously. In effect, most spikes fired by hippocampal neurons appear to be part of one sequence or another. Sequential organization may be fundamental, not only in the hippocampus, but also throughout the brain.

There is no tidy conclusion to this story; we are right in the middle of a second golden age of neurophysiology. Technical improvements in monitoring large numbers of cells simultaneously, and in distinguishing between previously hard to disentangle types of cells, are revolutionizing the field.[13] If history is anything to go by, what seem like fundamental mysteries today will yield to experimental results tomorrow. The brain is organized to depict sequences unfolding in time, generate imagery, and ponder alternative courses of action; it waits to tell us a story about itself that will be at once extraordinary and intuitive.

NOTES

1. The biggest simplification here is that the effect a neuron has on its downstream target is a function not only of whether or not it spiked, but also of the "strength" of the synaptic connection that it has to that target neuron, a variable that covers where the synapse is located on the target neuron, as well as biochemical and mechanical properties of the synapse. It is important that these properties can change, making the connection stronger or weaker; such change is thought to be how learning happens. It is this all-or-none property of neuronal spikes that prompted Warren McCulloch and Walter Pitts to suggest that neurons could act as logic gates, performing elementary logical functions on their inputs, and indeed their model inspired Von Neumann's design for the modern computer. W. S. McCulloch and W. Pitts, "A Logical Calculus of the Ideas Immanent in Nervous Activity," *Bulletin of Mathematical Biophysics* 5 (1943): 115.

2. Walter Pitts, also an author on this study (and therefore one of the most consequential figures in our story), was said to be so depressed by its challenge to his conception of neurons performing propositional logic that he was driven to alcoholism and ultimately an early death. Pitts was discovered by his future colleagues as a homeless runaway—the original "Good Will Hunting"—and his amazing story is recounted in a wonderful recent article: Amanda Gefter, "The Man Who Tried to Redeem the World with Logic," *Nautilus*, issue 021. Another great read on the subject is Jerome Lettvin's chapter in *Talking Nets: An Oral History of Neural Networks*, ed. James A. Anderson and Edward Rosenfeld (Cambridge: MIT Press, 1998).

3. You can see and hear the experimental observation of a simple cell—a cell that responds to an oriented line—at https://www.youtube.com/watch?v=Cw5PKV9Rj30.

4. H. B. Barlow, "Single Units and Sensation: A Neuron Doctrine for Perceptual Psychology," *Perception* 1 (1972): 317–394.

5. D. Marr, *Vision* (San Francisco: Freeman, 1982), ch. 1, "The Philosophy and the Approach," p. 15. It is curious that it has since been reported that single neurons recorded from the human brain can respond to the concept of a particular celebrity—for example, the "Jennifer Aniston cell"—whether delivered by image or the sound of the celebrity's name. R. Q. Quiroga, L. Reddy, G. Kreiman, C. Koch, and I. Fried, "Invariant Visual Representation by Single Neurons in the Human Brain," *Nature* 435 (2005): 1102–1107.

6. An accessible review by a key figure in the development of deep learning is the cheekily titled "Machines Who Learn," by Yoshua Bengio, *Scientific American* 314 (2016): 46–51. It is curious that while McCulloch-Pitts's logic neurons had little influence over subsequent neurophysiology, the simplified model of neural spiking they proposed has been the most influential model in neural networks, including deep learning, albeit the model has been put to different uses than the logical processing that they prescribed.

7. A superb description of the study of the most famous patient to suffer hippocampal damage is S. Corkin, *Permanent Present Tense: The Unforgettable Life of the Amnesic Patient, H. M.* (New York: Basic Books, 2013).

8. The classic original text that lays out the cognitive map hypothesis is freely available online: http://www.cognitivemap.net. Original citation: J. O'Keefe and L. Nadel, *The Hippocampus as a Cognitive Map* (Oxford: Clarendon Press, 1978). A more recent update that explains the later findings on grid cells and other kinds of spatial cells is E. I. Moser, E. Kropff, and M. B. Moser, "Place Cells, Grid Cells and the Brain's Spatial Representation System," *Annual Reviews in Neuroscience* 31 (2008): 69–89.

9. Ibid.

10. W. Singer, "Synchronization of Cortical Activity and Its Putative Role in Information Processing and Learning," *Annual Reviews in Physiology* 55 (1993): 349–374.

11. R. S. Zemel, P. Dayan, and A. Pouget, "Probabilistic Interpretation of Population Codes," *Neural Computation* 10 (1998): 403–430.

12. Work in any scientific field is a product of the ongoing efforts of many laboratories, and here are two recent reviews from laboratories that help integrate a lot of recent discoveries into a coherent story. The first considers how finding long sequences during the awake state changed our interpretation of their function, and the second considers the short "look-ahead" sequences that occur during movement: M. F. Carr, S. P. Jadhav, and L. M. Frank, "Hippocampal Replay in the Awake State: A Potential Substrate for Memory Consolidation and Retrieval," *Nature Neuroscience* 14 (2011): 147–153; A. M. Wikenheiser and A. D. Redish, "Decoding the Cognitive Map: Ensemble Hippocampal Sequences and Decision Making," *Current Opinion in Neurobiology* 32 (2015): 8–15. Some of the relevant primary sources from our laboratory are the following: B. E. Pfeiffer and D. J. Foster, "Hippocampal Place-Cell Sequences Depict Future Paths to Remembered Places," *Nature* 497 (2013): 74–79; X. Wu and D. J. Foster, "Hippocampal Replay Captures the Unique Topological Structure of a Novel Environment," *Journal of Neuroscience* 34 (2014): 6459–6469; D. Silva, T. Feng, and D. J. Foster, "Trajectory Events across Hippocampal Place Cells Require Previous Experience," *Nature Neuroscience* 18 (2015): 1772–1779.

13. Miniaturization has enabled the simultaneous electrophysiological recording of 263 neurons with place responses in a freely moving rat. Optogenetics is a technique developed from the natural light sensitivity of certain molecules

that enables the labeling of certain neurons to allow their selective stimulation via implanted optical fibers delivering laser light. Among the many possible applications of this technology is the identification of different neural cell types during electrophysiological recording.

A Comparative Approach Is Imperative for the Understanding of Brain Function

Cynthia F. Moss

HOW DOES THE BRAIN WORK? More specifically, how does the brain process and represent information from our surroundings? Does it generate commands for actions? Acquire and store new information? Recall past events? These are questions that capture the imagination. Rarely, however, does one ask the question, "What do we mean by THE brain?" Most people would probably respond, "The human brain." However, research methods are limited in studies of human brains, and various experimental animals, with unique ways of living, play an important role in scientific discovery. To truly understand the biological foundations of cognition and behavior, researchers must study and compare brains across the animal kingdom.

In recent decades, valuable new methods have been developed that permit studies of brain activity in humans performing cognitive tasks;[1] however, the spatial and temporal resolution of these noninvasive methods drastically limits what we can learn about brain function. Moreover, many invasive methods used to study brain function, particularly those that make measurements at a cellular and molecular level (like inserting electrodes to record electrical signals of neurons), cannot be conducted on human subjects, and the vast majority of neuroscience research relies on nonhuman animal subjects, mostly rodents, with the goal of linking discoveries to human brain function.

Many neuroscientists share the view that there are common features across animal nervous systems that can be uncovered through careful investigation; however, we certainly cannot know which features are common if we study the brain of only one species. There exists a culture in neuroscience that encourages and rewards scientists for tackling research questions with a limited set of experimental animals, like mice and rats, out of tradition, convenience, or competition, not because they are most appropriate to answer a particular research question. In this context, it is noteworthy that many rodent-based studies omit species information in publication titles, conveying the naïve view that they are reporting findings on "the one true brain."

To understand how brains produce behaviors, it is essential to embark on comparative studies of different animals that evolved to perform diverse natural tasks and to carefully choose animal subjects that are well suited to address the specific research problems under investigation. Comparative studies can help us understand which features of brain systems at the cellular-molecular, physiological, and anatomical levels are common across animals and which are specific to particular species. The common features found across species can help us identify general principles of nervous system design and function.

Comparative approaches in neuroscience have been informed by a principle articulated by the Nobel laureate August Krogh in 1929: "For such a large number of problems there will be some animal of choice or a few such animals on which it can be most conveniently studied."[2] For example, experiments on visual processing in the horseshoe crab, chosen for its suitability to address specific research questions, have had a resounding impact on neuroscience. Horseshoe crabs use vision to detect changing light levels and to find mates. They have compound eyes, anatomically very different from the eyes of humans and other vertebrate animals, a factor that has allowed scientists to conduct groundbreaking experiments and opened the door to major discoveries. The compound eyes of horseshoe crabs are comprised of about one thousand ommatidia, each with a cornea, lens, and receptors that respond to light, just like the rods and cones of the human eye. The visual receptors of horseshoe crabs are about one hundred times larger than mammalian rods and cones, and therefore scientists could isolate and measure light-evoked activity in separate, neighboring visual cells. H. K. Hartline and colleagues found that light stimulation of one visual cell serves to shut down activity in neighboring cells.[3] Hartline's discovery of "lateral inhibition" in the horseshoe crab eye, work that was recognized by a Nobel Prize in 1967, prompted wider investigations that uncovered evidence of lateral inhibition in mammalian vision, touch, and hearing,[4] as well as insights to perceptual enhancement of edges at stimulus boundaries in human subjects.[5]

Comparative studies in neuroscience, guided by observations of natural animal behaviors, can help scientists focus their research investigations on functionally important neural processes. Barn owls, for example, are nocturnal predators that rely on hearing to find food; these animals perch high in trees and listen to the sounds generated by prey as they scurry through ground vegetation and leaf litter. Observations of barn owl prey capture led to detailed behavioral studies of sound localization accuracy in these animals. Barn owls take advantage of asymmetric ear placement, which creates sound level differences at the left and right ears, to localize sound in the vertical plane. To localize sound sources in the horizontal plane, barn owls calculate the difference in arrival time of sound at the two ears.[6] This observation motivated careful study of the owl auditory system, and it revealed that the computation of sound source direction is based on two important operations: delay lines and coincidence detectors. Delay lines are created by axons of different lengths that project systematically to neurons in different locations of an auditory brain stem structure called nucleus laminaris. Coincidence detectors are neurons in nucleus laminaris that fire only when they are synchronously activated by delay-line input for a specific inter-aural time difference that denotes a particular sound source direction. The anatomical arrangement of delay lines and coincidence detectors creates a map of sound source location in the barn owl's central auditory system.[7] Specializations for sound localization in this animal serve to highlight the importance of spatial and temporal coding of auditory information and inform a broad understanding of hearing in a wide range of species.

Another fascinating example of a specialized animal that sheds light on important principles of nervous system function is the star-nosed mole. This animal has twenty-two fleshy appendages surrounding its snout that it uses to detect and discriminate prey in marshy wetlands. The star-nosed mole's appendages are exquisitely sensitive touch organs, containing more than twenty-five thousand sensory receptors, known as Eimer's organs.[8] The two appendages along the midline provide particularly detailed information about objects, and for this reason, they have been compared with the central portion of the human retina called the fovea. It is noteworthy that the star-nosed mole moves its tactile appendages rapidly to discriminate prey, just the way we move our eyes to inspect objects in our surroundings. Over half of this animal's brain is devoted to processing touch information from its twenty-two appendages, even though the star surrounding its snout comprises only about 10 percent of the mole's body. Within the somatosensory cortex of the star-nosed mole, there is a magnified representation of the midline star appendages that carry the highest resolution information, similar to the expanded region of visual cortex representing the high-acuity fovea of humans.[9] The specializations of the star-nosed mole contribute valuable general knowledge about the physiology and anatomy of mammalian touch systems, and they serve to highlight two fundamental principles: the importance of movement to sensing and the magnification in the amount of brain tissue devoted to fine sensory discrimination.

Comparative studies of sensory systems that serve comparable functions under different environmental conditions can illuminate general principles of neural information processing. For example, primates rely heavily on vision to find food in daylight, whereas echolocating bats rely heavily on hearing to localize insect prey in the dark. Although light is continuously available in the environment for animals that use vision, echolocating bats must emit sounds that return echoes from objects in the environment.[10] Features of sonar echoes, such as intensity, frequency content, and arrival time, allow the bat to discriminate and localize objects with high accuracy in complete darkness.[11] On the surface, these two distal sensing systems, vision and echolocation, seem quite distinct. However, experimental data show striking similarities between primate vision and bat echolocation in the mechanisms of gaze control,[12] neural responses to stimulus location,[13] and the influence of spatial attention on neural activity patterns.[14] These findings suggest that we can develop a broader and deeper understanding of how brains process information from the environment through detailed comparisons of different sensory systems in a broad range of species.

When it comes to investigations of larger networks of neurons in

mammalian brain systems, Krogh's principle is sometimes lost or forgotten. Instead, neuroscientists tend to employ traditional laboratory animal models, rodents, without consideration of species differences. The
fact is that the brains of rats and mice are no more similar to human
brains than those of species that are typically viewed as exotic, such as
bats or star-nosed moles. Indeed, studies of spatial navigation in bats have
served to highlight the importance of comparative research to separate
specializations from generalizations.

Pioneering work on the mechanisms of spatial navigation, carried
out in rodents by John O'Keefe and colleagues, led to the discovery of
"place cells," neurons in the hippocampus that are most active when an
animal occupies a restricted region in its environment, and the locations
associated with neural firing are referred to as "place fields" (also see
essay by David Foster in this volume).[15] Once hippocampal place fields
are established in lit conditions, activity patterns persist in the dark, indicating that they play a role in spatial memory.[16] May-Britt Moser, Edvard
Moser, and colleagues characterized another population of neurons in
the rodent hippocampal formation that show periodic place fields; these
fields are regularly spaced on a triangular grid.[17] These "grid cells," found
in brain regions surrounding the hippocampus, are implicated in distance measurement and way-finding when navigating though the world.[18]
O'Keefe, May-Britt Moser, and Edvard Moser, recognized for their elegant
research in systems neuroscience, were awarded the 2014 Nobel Prize.

Place cells and grid cells have more recently been characterized in
other mammals, notably humans,[19] nonhuman primates,[20] and bats.[21]
Whereas the discovery of place cells and grid cells in several different
species speaks to general mechanisms, digging deeper into comparative
data raises questions about how these neurons acquire their properties.

A great deal of research has been devoted to understanding how place
cells and grid cells acquire their place-dependent activity. In rodents,
neural firing patterns of place cells and grid cells have been tied to the
phase of a brain wave, known as theta rhythm, that oscillates at 5–12 Hz.
This brain wave runs continuously in the hippocampus of rodents as they
explore the environment by locomotion.[22] Many researchers attribute this
continuous theta rhythm to the creation of place cell and grid cell firing
activity, and experimental tests in rodents provide compelling evidence
for a theta-rhythm model of space representation.[23] However, compara-

tive data challenge this model. Notably, theta brain rhythm in humans,[24] nonhuman primates,[25] and bats[26] does not run continuously but instead occurs in bouts. The absence of continuous theta rhythm in primates and bats therefore raises specific questions: Do different mechanisms operate in rodents and other mammals in the representation of space? Or is continuous theta rhythm not actually implicated in space representation in any species? It is important that this example leads us back to the more general question: If we study brain function in only one species, what do we learn about THE brain?

NOTES

1. C. Senior, T. Russell, and M. Gazzaniga, *Methods in Mind* (Cambridge, MA: MIT Press, 2009).
2. A. Krogh, "The Progress of Physiology, *American Journal of Physiology* 90 (1929): 243–251. Indeed, Krogh's principal has guided big discoveries in neuroscience. For example, accessibility of the giant axon of the squid allowed fundamental measurements of the electrical signals in nerve cells (Hodgkin and Huxley, Nobel Prize 1963), and the simple nervous system of the California sea hare, *Aplysia californica*, enabled critical advances on the basic mechanisms of learning and memory (Kandel, Nobel Prize 2000).
3. H. K. Hartline, H. G. Wagner, and F. Ratliff, "Inhibition in the Eye of Limulus," *Journal of General Psychology* 5 (1956): 651–671.
4. S. Yantis, *Sensation and Perception* (New York: Worth Publishers, 2014).
5. T. Cornsweet, *Visual Perception* (New York: Academic Press, 1970).
6. E. I. Knudsen, "The Hearing of the Barn Owl," *Scientific American* 245 (1981): 113–125; E. I. Knudsen and M. Konishi, "A Neural Map of Auditory Space in the Owl," *Science* 200 (1978): 795–797.
7. C. E. Carr, "Delay Line Models of Sound Localization in the Barn Owl," *American Zoologist* 33 (1993): 79–85.
8. T. Eimer, "Die Schnauze des Maulwurfs als Tastwerkzeug," *Archiv für Microscopie und Anatomie* 7 (1873): 181–201.
9. K. C. Catania, "A Nose That Looks Like a Hand and Acts Like an Eye: The Unusual Mechanosensory System of the Star-Nosed Mole," *Journal of Comparative Physiology A* 185 (1999): 367–372; K. C. Catania, "Magnified Cortex in Star-Nosed Moles," *Nature* 375 (1995): 453–454.
10. D. R. Griffin, *Listening in the Dark* (New Haven: Yale University Press, 1958).
11. C. F. Moss and A. Surlykke, "Probing the Natural Scene by Echolocation," *Frontiers in Behavioral Neuroscience* 4 (2010): 33.
12. D. Sparks, "Translation of Sensory Signals into Commands for Control of Saccadic Eye Movements: Role of Primate Superior Colliculus," *Physiological Reviews* 66 (1986): 118–171; D. E. Valentine, S. Sinha, and C. F. Moss, "Orienting Responses and Vocalizations Produced by Microstimulation

of the Superior Colliculus of the Echolocating Bat," *Journal of Comparative Physiology* 188 (2002): 89–108.

13. R. H. Wurtz and J. E. Albano, "Visual-Motor Function of the Primate Superior Colliculus," *Annual Reviews in Neuroscience* 3 (1980): 189–226; D. E. Valentine and C. F. Moss, "Spatially-Selective Auditory Responses in the Superior Colliculus of the Echolocating Bat," *Journal of Neuroscience* 17 (1997): 1720–1733.

14. R. H. Wurtz and M. E. Goldberg, "The Primate Superior Colliculus and the Shift of Visual Attention," *Investigative Ophthalmology* 11 (1972): 441–450; R. M. McPeek and E. L. Keller, "Deficits in Saccade Target Selection after Inactivation of Superior Colliculus," *Nature Neuroscience* 7 (2004): 757–763.

15. J. O'Keefe, "Place Units in the Hippocampus of the Freely Moving Rat," *Experimental Neurology* 51 (1976): 78–109.

16. S. J. Y. Mizumori, *Hippocampal Place Fields* (Oxford: Oxford University Press, 2008).

17. M. Fyhn, S. Molden, M. P. Witter, E. I. Moser, and M.-B. Moser, "Spatial Representation in the Entorhinal Cortex," *Science* 305 (2004): 1258–1264; T. Hafting, M. Fyhn, S. Molden, M.-B. Moser, and E. I. Moser, "Microstructure of a Spatial Map in the Entorhinal Cortex," *Nature* 436 (2005): 801–806.

18. C. Barry, R. Hayman, N. Burgess, and K. J. Jeffery, "Experience-Dependent Rescaling of Entorhinal Grids," *Nature Neuroscience* 10 (2007): 682–684.

19. A. D. Ekström, M. J. Kahana, J. B. Caplan, T. A. Fields, E. A. Isham, E. L. Newman, and I. Fried, "Cellular Networks Underlying Human Spatial Navigation," *Nature* 425 (2003): 184–188.

20. N. J. Killan, M. J. Jutras, and E. A. Buffalo, "A Map of Visual Space in the Primate Entorhinal Cortex," *Nature* 491 (2012): 761–764.

21. N. Ulanovsky and C. F. Moss, "Hippocampal Cellular and Network Activity in Freely-Moving Echolocating Bats," *Nature Neuroscience* 10 (2007): 224–233; M. Yartsev and N. Ulanovsky, "Representation of Three-Dimensional Space in the Hippocampus of Flying Bats," *Science* 340 (2013): 367–372.

22. J. O'Keefe and M. L. Recce, "Phase Relationship between Hippocampal Place Units and the EEG Theta Rhythm," *Hippocampus* 3 (1993): 317–330; G. Buzsáki, "Theta Rhythm of Navigation: Link between Path Integration and Landmark Navigation, Episodic and Semantic Memory," *Hippocampus* 15 (2005): 827–840.

23. N. Burgess, C. Barry, and J. O'Keefe, "An Oscillatory Interference Model of Grid Cell Firing," *Hippocampus* 17 (2007): 801–912; M. E. Hasselmo, L. M. Giocomo, and E. A. Zilli, "Grid Cell Firing May Arise from Interference of Theta Frequency Membrane Potential Oscillations in Single Neurons," *Hippocampus* 17 (2007): 1252–1271.

24. A. D. Ekström, J. B. Caplan, E. Ho, K. Shattuck, I. Fried, and M. J. Kahana, "Human Hippocampal Theta Activity during Virtual Navigation," *Hippocampus* 15 (2005): 881–889.

25. M. J. Jutras, P. Fries, and E. A. Buffalo, "Oscillatory Activity in the Monkey

Hippocampus during Visual Exploration and Memory Formation," *Proceedings of the National Academy of Sciences of the USA* 6 (2013): 13144–13149.

26. M. M. Yartsev, M. P. Witter, and N. Ulanovsky, "Grid Cells without Theta Oscillations in the Entorhinal Cortex of Bats," *Nature* 479 (2011): 103–107.

The Cerebellum Learns to Predict the Physics of Our Movements

Scott T. Albert and Reza Shadmehr

TO INTERACT WITH and manipulate the world around us, our bodies must be constantly in motion. Even now, as you read the words on this page, your nervous system is coordinating a series of rapid eye movements (saccades) that allow your eyes to scan the text in front of you. These saccades are an example of a goal-directed movement; your brain chooses an interesting location in space (that is, the goal) and executes a movement to bring that spatial location into visual focus. Perhaps you might have a cup of coffee nearby. Periodically, your thirst (or your desire for a dose of caffeine) might drive you to interrupt your reading and reach for your cup. This process of reaching, like moving your eyes, requires you to select a target in the world around you and move your arm toward that target. Though this process of reaching may require little conscious effort on your part, moving toward an object in the space around you requires many different neural calculations.[1]

Let us consider this reaching movement in greater detail. To grasp the cup, your brain must move your arm from its current position to the position of the cup. Therefore, to plan your movement, your brain first has to figure out the location of the cup in front of you and also the current position of your arm. This process of localization is a sensory process. In other words, it requires the brain to use sensory information in the form of vision and proprioception (the oft-forgotten sense that allows you to sense the position of your body without vision) to determine the current positions of the cup and your arm. With the starting and ending positions in hand, parietal regions of your cerebral cortex compute the path of the arm that connects these positions in space—the trajectory of the movement. After the trajectory is determined, your primary motor cortex and other associated pre-motor areas then carefully transform this sensory signal into a motor plan—namely, the patterns of muscle contraction that will move your arm along the desired path toward the coffee.

So the process of moving toward a goal begins with sensory measurements about yourself and the world around you that are then converted into a set of motor actions. In short, the act of moving is guided by a sensation-to-movement map. This type of map is only one component of movement planning. It is intriguing that to execute our movements correctly, this transformation also runs in reverse—our brain uses planned motor actions to predict how the sensory state of our body will change should these actions be performed.

Why does the brain need to predict sensory events that might happen in the future? To answer this question, let us try an experiment. Take a book and place it in your left hand, and then ask a friend to pick up the book from your hand. You will notice that as the book is lifted off your hand, your hand does not stay perfectly still but shifts upward. Now place the book back in your left hand and use your right hand to pick up the book. Something remarkable happens: the left hand that was holding the book remains perfectly still.

This experiment has a practical lesson: at a fancy party when the waiter comes to you with a tray of drinks, don't pick up the glass yourself. Rather, let the waiter pick up the glass and give it to you. It also illustrates two fundamental ideas about our nervous system. First, sensory information (vision and proprioception) is acquired too slowly to allow us to maintain good control of our body. Despite the fact that we can see our friend reaching to pick up the book off our hand, this visual information arrives at the motor planning regions of our brain too late to be used to precisely reduce the muscle activity that was required to support the weight of the book. As a result of this delay, we overcompensate for the weight of the (now absent) book, and our hand shifts upward. Second, when we generate a movement (lift the book with our right hand), our brain predicts the sensory consequences of that movement before the movement occurs. As a result, when it is our own arm that is lifting the book, the brain predicts the sensory consequences that this action will have on our left arm and changes the muscle activity of our left arm to predictively compensate for the removal of the book. Predicting the sensory consequences of our motor commands allows us to overcome sensory delays that destabilize the control of our movements.[2]

We are all familiar with the consequences of incorrect planning of our motor actions: we screw up. In our coffee cup example, a problem in your brain's knowledge of how to reach to the cup in front of you will likely cause your hand to topple the cup, thus knocking its refreshing contents onto the table below. From our brain's perspective, the act of your knocking over the cup, instead of correctly picking it up for a sip, represents a mismatch between the predicted sensory state of our body and its actual sensory state. In neuroscience, we refer to this error as a sensory prediction error. It turns out that our brain continuously monitors these errors and uses them to fine-tune our motor behavior so that we do not make future mistakes. This process of error correction is so fundamental to our biology that distinct circuits of the brain appear to coordinate learning that results from error. One of these error correction pathways involves the cerebellum.[3] The cerebellum receives electrical information from a part of the brain stem called the inferior olive when a sensory prediction error occurs. The cerebellum uses these error signals to predict and correct for errors that might occur in the future.

Why is this process of error-based learning engrained in the physical structure of our nervous system? One answer to this question relates to the very nature of development. As we develop, our brain must learn to understand our bodies and our environment in order to solidify its sensation-to-movement maps that allow us to move accurately, quickly, and smoothly to the stimuli we sense around us. These sensation-to-movement maps require continual updating on many different timescales of life. Over long timescales, the brain must learn to change how it moves in response to the same stimulus because of changes to our physical bodies—for example, height, weight, and strength. Over short timescales, the brain must learn how to modify its motor actions because of muscle fatigue or perhaps changes in our environment such as movement through water versus movement on land. The inability to maintain and update sensation-to-movement maps significantly degrades our ability to make accurate movements. For example, individuals who suffer from certain cerebellar disorders (called cerebellar ataxias) exhibit extreme difficulties in the execution of nearly all voluntary movements such as reaching, talking, looking around, and walking. These deficits can, in some cases, confine an ataxic patient to a wheelchair.

Neuroscientists who study motor learning think that one key to understanding movement disorders like cerebellar ataxias lies in elucidating the process by which the brain updates its sensation-to-movement maps over time—in other words, how it uses sensory prediction errors to correct for movement errors in the future. Over many decades, the motor learning community has devised clever experimental protocols that allow researchers to carefully control errors that an individual experiences and precisely measure how individuals correct their movements in the future. One of these protocols, known as reach adaptation, focuses specifically on how people learn to correct reaching movements of their arms or, in more prosaic terms, how people avoid knocking over cups of coffee.

To study how people adapt their reaching movements over time, we ask participants to reach from one point to another while holding the handle of a robotic arm. As the subject makes a reach, the experimenter can perturb the movement of the subject's arm by imposing forces on the hand via the robot. In other words, by pushing on the subject's hand, we can convert normally straight and ideal movements into curved, distorted movements.[4] In order to reach the target in a specified amount of time, the subject must learn to produce additional forces that predict and counteract the robot's intervention. It is remarkable that healthy adults can almost completely eliminate the errors caused by the force field in just a few dozen trials.

How does the nervous system learn the appropriate corrections so rapidly? To answer this question, it is useful to consider how people learn things in everyday life: we use teachers. If we want to get better at basketball, we take basketball lessons. If we want to learn a new language, we seek a language instructor. We refer to a teacher-driven learning process as supervised learning. Is it possible that a supervised learning process is used for the correction of our movements? If so, who is the teacher? A good teacher is someone who already knows how to do the task we seek to learn. So for movement correction, a good teacher would be someone who knows how to eliminate a sensory error by changing our patterns of muscle activation. Fortunately, healthy individuals have such a teacher; it is called a reflex.

Our reflexes are part of our automatic error correction circuit, and they naturally coordinate the sensation-to-movement map we seek: they observe a sensory error and rapidly correct it with a motor response. For example, when we slip, our sensory imbalance causes us to throw our hands out in front of us without thinking so as to catch our imminent fall. Everyone is likely familiar with the stretch reflex. When a physician abruptly hits the patellar ligament of the knee with a dense object, our quadriceps muscle stretches rapidly, and our spinal reflexes quickly counteract this unintended stretch through a reciprocal contraction that causes us to kick our leg out in front of us. These stretch reflexes are also present in our arms and can be measured in the laboratory using electromyography, which is a technique for recording muscle activity to precisely measure the muscle signals that move our arms in space during both voluntary movement and subconscious reflexive movement.

Let us now return to our question regarding force field adaptation. Is it possible that the reflexes that counteract the unexpected force field also instruct the brain on how to produce better motor commands in the future? Preliminary evidence suggests that the answer is yes. The electromyelogram patterns that our reflexes add to our movement during a force field perturbation closely resemble the changes we make to our future muscle activation patterns. In other words, corrective muscle activity during a single movement appears to instruct future changes in our motor actions on the next reach.[5] For example, if we contract our biceps and relax our triceps in response to the imposing force field, our motor system appears to incorporate this feedback correction into a new sensation-to-movement map—that is, the brain thinks that to achieve the desired straight-line movement in the future, the biceps needs to be more active and the triceps needs to be less active than they were during the original movement.

To summarize, the movements we make are a bit more complicated than they might seem at first glance; they involve complex conversions between sensory predictions and motor actions. And the conversions themselves are constantly being tweaked as we learn more about ourselves and our environment. This learning is triggered by sensory prediction errors, or mismatches between what we expected to feel after moving and what we actually feel. Here we have explored this process of learning and culminated with the idea that to improve our future movements, our brain likely seeks advice from teaching systems embedded within our brain and spinal cord—namely, our reflexes.

Insights into the control of movement, as we have discussed here, can have large implications in the understanding of our behavior in both healthy and diseased populations. For example, our discussion may inform the question of why some individuals learn motor skills faster than others; perhaps they have a better system of reflexes. More important, perhaps the motor deficits that accompany various movement disorders, such as stroke and cerebellar ataxia, are in part caused by patients' inability to execute appropriate corrections to their movements, thus depriving their brains of an extremely knowledgeable teacher. This research suggests that encouraging patients to make mistakes while moving and reinforcing patient-driven feedback corrections to their movement errors may be one path toward neurorehabilitation. In other words, as neuroscientists, we hope that the fundamental insights we make concerning the brain will be translated into methods for improving human life. At the very least, hopefully in the not-so-distant future, we will never spill our coffee again.

NOTES

1. R. Shadmehr and S. P. Wise, *Computational Neurobiology of Reaching and Pointing: A Foundation for Motor Learning* (Cambridge, MA: MIT Press, 2005).
2. R. Shadmehr, M. A. Smith, and J. W. Krakauer, "Error Correction, Sensory Prediction, and Adaptation in Motor Control," *Annual Reviews of Neuroscience* 33 (2010): 89–108.
3. The cerebellum (Latin for "little brain") is a brain region located at the back of the head that contains roughly half of all neurons in our brain. R. Shadmehr and J. W. Krakauer, "A Computational Neuroanatomy for Motor Control," *Experimental Brain Research* 185 (2008): 359–381.
4. A common way to perturb a subject's movement is by applying a force field to the arm; one of the most common force fields used in reach adaptation is called a curl field. In a curl field, the robot will push the subject's arm in a direction that is perpendicular to the intended direction of motion. R. Shadmehr, "Learning to Predict the Physics of Our Movements," *Journal of Neuroscience* 37 (2016): 1663–1671.
5. S. Albert and R. Shadmehr, "The Neural Feedback Response to Error as a Teaching Signal for the Motor Learning System," *Journal of Neuroscience* 36 (2016): 4832–4845.

Neuroscience Can Show Us a New Way to Rehabilitate Brain Injury

THE CASE OF STROKE

John W. Krakauer

HISTORICALLY, awareness of stroke can be traced back to the time of the Edwin Smith Papyrus, which was written in Egypt about 1600 BCE.[1] Almost a millennium later on the Indian subcontinent, the Vedic term *pakshavhada* was coined to refer to hemiplegia (paralysis down one half of the body).[2] Stroke, however, was first written about formally by Hippocrates around 400 BCE and was referred to in Greek as *apoplexia* (αποπληξία), which means to suddenly strike to earth, to knock down with violence.[3] The word "stroke" itself is thought to derive from an Old English word, *strac*, a hard blow.[4] Thus the etymology emphasizes the devastating out-of-the-blue quality of stroke whereby a person is fine one minute and disabled the next. Indeed, Hippocrates, who was particularly interested in prognosis, stated that *"it is impossible to remove a strong attack of apoplexia, and not easy to remove a weak attack."*[5] To this day we are still faced with the very daunting challenge of trying to "remove" the effects of stroke after they have happened.

Stroke is a leading cause of physical adult disability worldwide.[6] Approximately 795,000 people experience either a new or recurrent stroke in the United States each year; it means that someone is having a stroke every forty seconds.[7] The great majority, almost nine out of every ten, are ischemic strokes.[8] Ischemic stroke refers to the death of an area of brain tissue (infarction) due to loss of its supply of oxygen (ischemia) when, in most cases, the artery supplying it with blood is blocked by a clot or an atherosclerotic plaque. This essay focuses on arm and hand paresis after stroke, but the principles introduced almost certainly apply to a great extent to other deficits, including leg paresis and loss of language (aphasia). Clinicians use the term "paresis" (rather than "paralysis") to capture both limb weakness and loss of motor control. The term "hemiparesis" is also frequently used because the most common poststroke deficit is paresis in the face, arm, and leg on the opposite side of the body from the cerebral hemisphere in which the stroke occurred.

It is important to appreciate that loss of control and strength after stroke are distinct deficits and can dissociate: it is not uncommon for a patient after stroke to be able to squeeze quite hard with all fingers (power grip) but be unable to move any given finger independently of the others (individuation).[9] Analogously for the arm, a patient may be able to lift it at the shoulder but be unable to point accurately.[10] That the brain might have different substrates for control and force can be intuited by considering that an Olympic weightlifter might have a much stronger grip than a concert pianist but not have the same degree of finger dexterity. About 75 percent of patients with acute stroke will have arm or hand paresis; over 60 percent of these will still have a deficit six months after the stroke that continues to interfere with function.[11]

At the current time no form of rehabilitation has meaningful impact on poststroke impairment of the arm or hand. That is, no treatment reverses either the motor control deficits or weakness caused by the stroke. Indeed, perhaps in tacit admission of this fact, which has been recognized since the time of Hippocrates, the entire system of modern rehabilitative care after stroke focuses instead on relearning activities of daily living, like dressing and eating, at a *fixed* level of impairment. The main concept is that of *compensation,* which is the use of motor learning principles and task-specific training to exploit the capacities that a patient has left—so, for example, use of the left arm for daily activities if the right arm is no longer able. Another example of compensation but on the affected side is to move the upper trunk forward to aid in a reach if the elbow can no longer be fully extended. Compensation can be contrasted with *restitution,* which is the attempt to promote brain repair and a return to premorbid patterns of motor control.[12]

Modern rehabilitation of stroke therefore consists of training on real-life tasks so as to reduce disability and promote independence, with a concomitant emphasis on compensatory strategies rather than restitution. This approach, for better or worse, has also promoted shorter lengths of stay in inpatient rehabilitation units.[13] It would be reassuring if this present state of neurorehabilitation could be attributed to scientific considerations, but unfortunately high-quality evidence is lacking for any of the interventions now given on rehabilitation units;[14] instead current practice has largely been dictated by economics. There have been calls to subject current rehabilitation approaches to proper investigation,[15] but here the case will be made that we should not be trying to eke out any more gains from existing therapies or to study them further. An entirely different approach is needed, especially in the first weeks to months after stroke, the period when most rehabilitation is given. The goal needs to be to reverse impairment—a goal that, as we have seen, has been considered out of reach for millennia. To understand why impairment may indeed be amenable to therapeutic intervention it is first necessary to introduce the notion of *spontaneous biological recovery*.

Spontaneous biological recovery refers to the phenomenon that almost all patients recover from impairment to some degree after stroke and that almost all of this recovery occurs in the first three months.[16] The word "spontaneous" is somewhat controversial because it implies that recovery is attributable to an endogenous repair process rather than current rehabilitative interventions. There is, however, good evidence to support this claim. In 2008, we showed that recovery from motor impairment in the arm in the first three months after stroke could be captured by a surprisingly simple and predictable rule, which we called *proportional recovery*. The rule states that most patients will recover a fixed 70 percent proportion of their maximal potential recovery at three months.[17] The fact that this proportional rule exists at all—a rule that has since been reproduced several times[18]—implies that existing therapies cannot be having much of an additional impact above that expected from spontaneous biological recovery. How does this follow? The answer lies in the fact that maximal potential recovery at three months is computed from the level of impairment in the first week after stroke. For initial impairment at one week to predict so well where a patient can expect to be at three months means that conventional therapy in between these two time points cannot be appreciably impinging upon this relationship. This has since been shown to be the case empirically: extra conventional therapy does not alter the proportional recovery rule.[19] Finally, it has recently been shown that recovery of motor control using careful measurement of movement plateaus at about five weeks after stroke, despite continued recovery of strength and activity.[20]

To summarize thus far, early after a stroke two phenomena seem to be going on, and they are passing each other like ships in the night: conventional therapy targets activities of daily living by training compensatory strategies, while in parallel spontaneous biological recovery reverses impairment. Is there a way to move beyond this impasse and somehow target impairment—that is, can we develop rehabilitation approaches that piggyback on spontaneous biological recovery mechanisms and allow patients to recover beyond 70 percent? A clue to this question comes from studies using stroke induced in rodents in the laboratory. There is a surprising similarity between rodents and humans when it comes to prehension (reach and grasp).[21] If a motor cortical stroke is induced in a rat or a mouse after the animal has been trained to maximal levels of performance on a prehension task, it recovers much better if it is rehabilitated with retraining on the task earlier than later.[22] For example, if a mouse is trained on a prehension task in which it must reach through a narrow slit to pick up a food pellet with its paw, it recovers back to normal prestroke levels if it starts training a day after the stroke but only minimally if retraining is delayed by a week.[23]

The results in rodents suggest that there is something about ischemic stroke itself that induces a time-limited window of augmented responsiveness to training. Dramatic proof of this conjecture came from a recent experiment by Steve Zeiler and colleagues at Johns Hopkins University School of Medicine.[24] They reasoned that a second motor cortical stroke might paradoxically reopen a sensitive period of responsiveness to training and promote full recovery from a previous first stroke. To test this they gave mice a first stroke in motor cortex and then waited a week before beginning retraining. As expected, the mice recovered only minimally because too much time had been allowed to pass before training was initiated. They then gave these same mice a second stroke in an area near to the original stroke, and, not surprisingly, the animals developed an even worse impairment. The surprising result was that with retraining the mice returned to normal levels of performance. In essence a previous stroke was treated with a new stroke. It should be made clear that this experiment was done to prove definitively that there is a sensitive period after stroke that allows training to promote full recovery at the level of impairment.

It is clearly not a viable therapeutic option to induce a second stroke in patients after a first stroke. Other means will need to be found to have the same desired effect without causing more damage to the brain. One promising option is to combine drugs, such as the serotonin reuptake inhibitor Fluoxetine (Prozac), with training early after stroke.[25] Another is to drastically increase the intensity and dosage of behavioral training that patients receive early after stroke. At the current time in the first weeks after stroke patients spend about 60 percent of their time alone and 85 percent of the time immobile.[26] We know from basic science that hundreds, if not thousands, of movement repetitions are needed to induce detectable changes in motor cortex in animal models.[27] Current therapy offers only about thirty![28]

The studies mentioned here have show that there is early (first three months) spontaneous recovery after stroke in humans and a short-lived window (one week) of increased responsiveness to training in experimental rodent strokes. How can this information be combined to develop new rehabilitation approaches that go beyond task-specific compensatory training? First, one must begin with the assumption that there is mechanistic overlap between spontaneous recovery in humans and the sensitive period shown in mice. If this assumption is correct, then it is to be hoped that a window of increased responsiveness to training exists in humans and that it may last up to three months. A second assumption is that it should be possible to devise a form of motor training that can be given at sufficiently high dose within the sensitive period so that it leads to more general gains rather than just task-specific ones—that is, it overcomes motor learning's curse of task specificity.[29] For example, being a fast typist does not make you a fast shoe lace tier, even though you use your fingers in both cases. One reason for the assumption that such task specificity may be overcome is that there are some similarities between the plastic state induced by stroke and normal early development.[30]

Based on the neuroscience I have just described and the attendant assumptions, my lab has developed a customized video game based on a highly immersive virtual oceanic environment in which a patient controls a simulated dolphin that responds to the patient's desired steering direction in the water.[31] The dolphin has to jump and spin, catch fish, and battle sharks.[32] The game was designed to promote intrinsic motivation, which promotes more enjoyment, creativity, and creative flexibility than motivation obtained from external rewards.[33] The dolphin game was designed to be used in combination with an exoskeletal robot arm for an ongoing multicenter trial that we are conducting for patients with moderate to severe hemiparesis after stroke. The goal is to have the patient make a large number of exploratory arm movements, somewhat analogous to the playful nongoal-directed arm movements that infants make (motor babbling), for two hours a day for three weeks.

The exoskeletal robot provides antigravity support so that the patients can practice at control despite their weakness. To be eligible for the trial the patients cannot be more than five weeks out from stroke onset. The core hypothesis is that general recovery from impairment above what is expected from spontaneous mechanisms will occur because patients will perform miles of continuous arm movements per day that cover the space of arm configurations that are used across a range of everyday tasks[34] and will do so within a poststroke-sensitive period. Skilled exploratory movements at this dose and intensity have never been tried before so early after stroke. It is an empirical question whether results as dramatic as have been seen in animal models will translate to patients. Regardless, it is to be hoped that the reasoning and work presented here, which culminated in the testing of a new rehabilitation approach, exemplify the potential of translating ideas and findings from neuroscience to human health and well-being.

NOTES

1. J. H. Breasted, *The Edwin Smith Surgical Papyrus: Published in Facsimile and Hieroglyphic Transliteration with Translation and Commentary in Two Volumes* (Chicago: University of Chicago Press, 1930).
2. S. Mishra, B. Trikamji, S. Singh, P. Singh, and R. Nair, "Historical Perspective of Indian Neurology," *Annals of Indian Academy of Neurology* 16 (2013): 467–477.
3. E. Clarke, "Apoplexy in the Hippocratic Writings," *Bulletin of the History of Medicine* 37 (1963): 301–314.
4. R. K. Barnhart and S. Steinmetz, *The Barnhart Dictionary of Etymology* (Bronx: H. W. Wilson, 1987).

5. Quoted in J. A. Lidell, *A Treatise on Apoplexy, Cerebral Hemorrhage, Cerebral Embolism, Cerebral Gout, Cerebral Rheumatism, and Epidemic Cerebro-spinal Meningitis* (New York: W. Wood, 1873), p. 14.

6. Centers for Disease Control and Prevention, "Prevalence and Most Common Causes of Disability among Adults—United States, 2005," *MMWR: Morbidity and Mortality Weekly Report* 58 (2009): 421–426; V. L. Feigin et al., "Global and Regional Burden of Stroke during 1990–2010: Findings from the Global Burden of Disease Study 2010," *Lancet* 383 (2014): 245–254; World Health Organization, "Cardiovascular Diseases (CVDs)," http://www.who.int/mediacentre/factsheets/fs317/en/.

7. D. Mozaffarian et al., "Heart Disease and Stroke Statistics—2016 Update," *Circulation* 133, no. 4 (January 2016): 338–360.

8. Ibid.

9. P. Raghavan, E. Petra, J. W. Krakauer, and A. M. Gordon, "Patterns of Impairment in Digit Independence after Subcortical Stroke," *Journal of Neurophysiology* 95 (2006): 369–378.

10. K. M. Zackowski, A. W. Dromerick, S. A. Sahrmann, W. T. Thach, and A. J. Bastian, "How Do Strength, Sensation, Spasticity and Joint Individuation Relate to the Reaching Deficits of People with Chronic Hemiparesis?" *Brain* 127 (2004): 1035–1046.

11. J. Bamford, P. Sandercock, L. Jones, and C. Warlow, "The Natural History of Lacunar Infarction: The Oxfordshire Community Stroke Project," *Stroke* 18 (1987): 545–551; D. T. Wade and R. L. Hewer, "Functional Abilities after Stroke: Measurement, Natural History and Prognosis," *Journal of Neurology, Neurosurgery and Psychiatry* 50 (1987): 177–182; R. Bonita and R. Beaglehole, "Recovery of Motor Function after Stroke," *Stroke* 19 (1988): 1497–1500; A. Sunderland, D. Tinson, L. Bradley, and R. L. Hewer, "Arm Function after Stroke: An Evaluation of Grip Strength as a Measure of Recovery and a Prognostic Indicator," *Journal of Neurology, Neurosurgery and Psychiatry* 52 (1989): 1267–1272; S. S. Rathore, A. R. Hinn, L. S. Cooper, H. A. Tyroler, and W. D. Rosamond, "Characterization of Incident Stroke Signs and Symptoms: Findings from the Atherosclerosis Risk in Communities Study," *Stroke* 33 (2002): 2718–2721.

12. J. W. Krakauer, S. T. Carmichael, D. Corbett, and G. F. Wittenberg, "Getting Neurorehabilitation Right: What Can Be Learned from Animal Models," *Neurorehabilitation and Neural Repair* 26 (2012): 923–931; M. F. Levin, J. A. Kleim, and S. L. Wolf, "What Do Motor 'Recovery' and 'Compensation' Mean in Patients Following Stroke?" *Neurorehabilitation and Neural Repair* 23 (2009): 313–319; A. Roby-Brami, A. Feydy, M. Combeaud, E. V. Biryukova, B. Bussel, and M. F. Levin, "Motor Compensation and Recovery for Reaching in Stroke Patients," *Acta Neurologica Scandinavica* 107 (2003): 369–381; P. Raghavan, M. Santello, A. M. Gordon, and J. W. Krakauer, "Compensatory Motor Control after Stroke: An Alternative Joint Strategy for Object-Dependent Shaping of Hand Posture," *Journal of Neurophysiology* 103 (2010): 3034–3043.

13. M. Camicia, H. Wang, M. DiVita, J. Mix, and P. Niewczyk, "Length of Stay at Inpatient Rehabilitation Facility and Stroke Patient Outcomes," *Rehabilitation Nursing* 41 (2016): 78–90.

14. A. Pollock, S. E. Farmer, M. C. Brady, P. Langhorne, G. E. Mead, J. Mehrholz, and F. van Wijck, "Interventions for Improving Upper Limb Function after Stroke," *Cochrane Database of Systematic Reviews* 12 (2014): CD010820.

15. Ibid.

16. Krakauer, Carmichael, Corbett, and Wittenberg, "Getting Neurorehabilitation Right"; S. C. Cramer, "Repairing the Human Brain after Stroke: I. Mechanisms of Spontaneous Recovery," *Annals of Neurology* 63 (2008): 272–287; P. W. Duncan, L. B. Goldstein, D. Matchar, G. W. Divine, and J. Feussner, "Measurement of Motor Recovery after Stroke: Outcome Assessment and Sample Size Requirements," *Stroke* 23 (1992): 1084–1089; H. Nakayama, H. S. Jørgensen, H. O. Raaschou, and T. S. Olsen, "Recovery of Upper Extremity Function in Stroke Patients: The Copenhagen Stroke Study," *Archives of Physical Medicine and Rehabilitation* 75 (1994): 394–398; H. S. Jørgensen, H. Nakayama, H. O. Raaschou, and T. S. Olsen, "Stroke. Neurologic and Functional Recovery: The Copenhagen Stroke Study," *Physical Medicine and Rehabilitation Clinics of North America* 10 (1999): 887–906.

17. S. Prabhakaran, E. Zarahn, C. Riley, A. Speizer, J. Y. Chong, R. M. Lazar, R. S. Marshall, and J. W. Krakauer, "Inter-Individual Variability in the Capacity for Motor Recovery after Ischemic Stroke," *Neurorehabilitation and Neural Repair* 22 (2008): 64–71.

18. E. R. Buch, S. Rizk, P. Nicolo, L. G. Cohen, A. Schnider, and A. G. Guggisberg, "Predicting Motor Improvement after Stroke with Clinical Assessment and Diffusion Tensor Imaging," *Neurology* 86 (2016): 1924–1925; C. Winters, E. E. H. v. Wegen, A. Daffertshofer, and G. Kwakkel, "Generalizability of the Proportional Recovery Model for the Upper Extremity after an Ischemic Stroke," *Neurorehabilitation and Neural Repair* 29 (2015): 614–622.

19. W. D. Byblow, C. M. Stinear, P. A. Barber, M. A. Petoe, and S. J. Ackerley, "Proportional Recovery after Stroke Depends on Corticomotor Integrity," *Annals of Neurology* 78 (2015): 848–859.

20. J. C. Cortes, J. Goldsmith, M. D. Harran, J. Xu, N. Kim, H. M. Schambra, A. R. Luft, P. Celnik, J. W. Krakauer, and T. Kitago, "A Short and Distinct Time Window for Recovery of Arm Motor Control Early after Stroke Revealed with a Global Measure of Trajectory Kinematics," *Neurorehabilitation and Neural Repair* 31 (2017): 552–560.

21. I. Q. Whishaw, S. M. Pellis, and B. P. Gorny, "Skilled Reaching in Rats and Humans: Evidence for Parallel Development or Homology," *Behavioural Brain Research* 47 (1992): 59–70; L.-A. R. Sacrey, M. Alaverdashvili, and I. Q. Whishaw, "Similar Hand Shaping in Reaching-for-Food (Skilled Reaching) in Rats and Humans Provides Evidence of Homology in Release, Collection, and Manipulation Movements," *Behavioural Brain Research* 204 (2009): 153–161.

22. J. Biernaskie, G. Chernenko, and D. Corbett, "Efficacy of Rehabilitative

Experience Declines with Time after Focal Ischemic Brain Injury," *Journal of Neuroscience* 24 (2004): 1245–1254; K. L. Ng, E. M. Gibson, R. Hubbard, J. Yang, B. Caffo, R. J. O'Brien, J. W. Krakauer, and S. R. Zeiler, "Fluoxetine Maintains a State of Heightened Responsiveness to Motor Training Early after Stroke in a Mouse Model," *Stroke* 46 (2015): 2951–2960; S. R. Zeiler and J. W. Krakauer, "The Interaction between Training and Plasticity in the Poststroke Brain," *Current Opinion in Neurology* 26 (2013): 609–616.

23. Ng, Gibson, Hubbard, Yang, Caffo, O'Brien, Krakauer, and Zeiler, "Fluoxetine Maintains a State of Heightened Responsiveness to Motor Training Early after Stroke in a Mouse Model."

24. S. R. Zeiler, R. Hubbard, E. M. Gibson, T. Zheng, K. Ng, R. O'Brien, and J. W. Krakauer, "Paradoxical Motor Recovery from a First Stroke after Induction of a Second Stroke Reopening a Postischemic Sensitive Period," *Neurorehabilitation and Neural Repair* 30 (2016): 794–800.

25. Ng, Gibson, Hubbard, Yang, Caffo, O'Brien, Krakauer, and Zeiler, "Fluoxetine Maintains a State of Heightened Responsiveness to Motor Training Early after Stroke in a Mouse Model"; V. Windle and D. Corbett, "Fluoxetine and Recovery of Motor Function after Focal Ischemia in Rats," *Brain Research* 1044 (2005): 25–32; F. Chollet, J. Tardy, J. F. Albucher, C. Thalamas, E. Berard, C. Lamy, Y. Bejot, S. Deltour, A. Jaillard, P. Niclot, B. Guillon, T. Moulin, P. Marque, J. Pariente, C. Arnaud, and I. Loubinoux, "Fluoxetine for Motor Recovery after Acute Ischaemic Stroke (FLAME): A Randomised Placebo-Controlled Trial," *Lancet Neurology* 10 (2011): 123–130.

26. J. Bernhardt, H. Dewey, A. Thrift, and G. Donnan, "Inactive and Alone: Physical Activity within the First 14 Days of Acute Stroke Unit Care," *Stroke* 35 (2004): 1005–1009.

27. R. J. Nudo, G. W. Milliken, W. M. Jenkins, and M. M. Merzenich, "Use-Dependent Alterations of Movement Representations in Primary Motor Cortex of Adult Squirrel Monkeys," *Journal of Neuroscience* 16 (1996): 785–807; E. J. Plautz, G. W. Milliken, and R. J. Nudo, "Effects of Repetitive Motor Training on Movement Representations in Adult Squirrel Monkeys: Role of Use versus Learning," *Neurobiology of Learning and Memory* 74 (2000): 27–55; K. M. Friel, S. Barbay, S. B. Frost, E. J. Plautz, A. M. Stowe, N. Dancause, E. V. Zoubina, and R. J. Nudo, "Effects of a Rostral Motor Cortex Lesion on Primary Motor Cortex Hand Representation Topography in Primates," *Neurorehabilitation and Neural Repair* 21 (2007): 51–61.

28. C. E. Lang, J. R. Macdonald, D. S. Reisman, L. Boyd, T. Jacobson Kimberley, S. M. Schindler-Ivens, T. G. Hornby, S. A. Ross, and P. L. Scheets, "Observation of Amounts of Movement Practice Provided during Stroke Rehabilitation," *Archives of Physical Medicine and Rehabilitation* 90 (2009): 1692–1698.

29. D. Bavelier, C. S. Green, A. Pouget, and P. Schrater, "Brain Plasticity through the Life Span: Learning to Learn and Action Video Games," *Annual Review of Neuroscience* 35 (2012): 391–416.

30. S. C. Cramer and M. Chopp, "Recovery Recapitulates Ontogeny," *Trends in Neurosciences* 23 (2000): 265–271.

31. K. Russell, "Helping Hand—Robots, Video Games, and a Radical New Approach to Treating Stroke Patients," Annals of Medicine, *New Yorker,* Nov. 23, 2015.

32. C. Zimmer, "Could Playing a Dolphin in a Video Game Help Stroke Patients Recover?" https://www.statnews.com/2016/03/25/after-a-stroke-reteaching -the-brain-by-gaming/. This article shows a video of the therapeutic game.

33. D. A. Gentile, "The Multiple Dimensions of Video Game Effects," *Child Development Perspectives* 5 (2011): 75–81; A. K. Przybylski, C. Scott, and R. M. Ryan, "A Motivational Model of Video Game Engagement," *Review of General Psychology* 14 (2010): 154–166; K. Laver, S. George, S. Thomas, J. E. Deutsch, and M. Crotty, "Virtual Reality for Stroke Rehabilitation: An Abridged Version of a Cochrane Review," *European Journal of Physical and Rehabilitation Medicine* 51 (2015): 497–506; C. S. Green and D. Bavelier, "Action Video Game Modifies Visual Selective Attention," *Nature* 423 (2003): 534–537; T. Strobach, P. A. Frensch, and T. Schubert, "Video Game Practice Optimizes Executive Control Skills in Dual-Task and Task Switching Situations," *Acta Psychol (Amst)* 140 (2012): 13–24.

34. I. S. Howard, J. N. Ingram, K. P. Kording, and D. M. Wolpert, "Statistics of Natural Movements Are Reflected in Motor Errors," *Journal of Neurophysiology* 102 (2009): 1902–1910.

Almost Everything You Do Is a Habit

Adrian M. Haith

DO YOU CHECK your smartphone every few minutes? Maybe you bite your nails or eat too much chocolate. Or perhaps you go to the gym regularly and always wash up straight after dinner.

Habits, good or bad, are a fact of life. We all perform some actions habitually without actually thinking about them. Sometimes these actions are undesirable, and we resent their intrusion into our otherwise thoughtful and deliberate existence. Others exert a positive influence on our lives, and we deliberately cultivate them as a force for self-improvement. Our daily lives appear scattered with habits, but in truth we are aware of only a small minority of them. Beneath the surface of our cognition lurks a vast and unseen conglomeration of habits that drives almost everything we do.

For the purposes of scientific inquiry, the definition of "habit" as a thoughtlessly generated action is rather too loose. It is difficult to precisely define or measure thoughtlessness in humans, let alone in monkeys or rats. Neuroscientists have therefore adopted a definition of "habit" that is more precise and is measurable. The definition depends on the idea that a person or animal always has a goal. For example, your goal might be to drive to the grocery store to buy milk. In pursuit of this goal, every decision you make or action you choose should ideally be one that brings you closer to this goal. If so, your behavior is said to be *goal-directed*.

By contrast, if your choice of action is the same regardless of your goal, then your behavior is *habitual*. Though your goal may be to pick up milk on the way home from work, you might nevertheless habitually drive straight home. It is important that the same habit can be either detrimental or beneficial. The habit of driving straight home from work is appropriate if you already have milk in the refrigerator. Critically, therefore, what distinguishes habitual from goal-directed behavior is not the suitability of your action choices but whether or not they adapt when your goals change.

This formal definition of a habit encompasses the usual behaviors we think of as habits and also helps to clarify the boundary between regular behavior and a habit. You may eat chocolate because you like the taste. But if you find that you eat chocolate even when you don't find it appetizing, then it has become a habit. It is of course sensible to check your phone periodically to ensure you haven't missed an important message. But if you find yourself doing it even when you are in a foreign country with no network service, then it has become a habit. There may be days when you really don't feel motivated to work out. But if you make it to the gym anyway, then going to the gym has become a habit.

The definition of "habit" in terms of inflexibility provides a simple means to test whether a given behavior is a habit: simply alter the goals of a task and see whether or not the behavior changes accordingly. This approach can be used to test what is and what is not a habit, as well as to study how habits are formed in the first place. A classic example is to place a hungry rat in a maze containing food at a particular location, day after day. Within a few days, the rat starts to behave in a stereotypical way, always following the same path to the food. In order to test whether this behavior is a habit or not, the experimenter might one day send the rat in with a full stomach but deprived of water. In this case, if acting habitually, the thirsty rat will run directly to the food source and then not eat a bite, whereas a goal-directed rat will be more circumspect and explore other options.

Experiments along these lines have revealed how practice leads to a transition from goal-directed to habitual behavior. During the first few sessions performing a new task, the rat exhibits behavior that is goal-directed; the behavior changes whenever the rat's motivational state changes. After prolonged training, however, the rat eventually exhibits a

habitual pattern, persisting in the same rote behavior despite its not leading to something it wants.[1] This progression from goal-directed behavior
early in learning to habitual behavior after practice is a critical hallmark
of learning, whether in the context of navigating through a maze, learning new movement patterns, or more high-level behaviors like maintaining a healthy diet.[2] Therefore, while we tend to think of our daily behavior as rational and goal-directed, being goal-directed is in fact quite a
fleeting state, reserved for novel environments and circumstances. With
practice and repetition, goal-directed behavior is rapidly archived into a
habit, to be invoked whenever the appropriate circumstances arise.

Once you begin thinking of habits in terms of inflexibility rather
than utility, it does not take long to identify examples that illustrate just
how habitual our everyday lives are. Consider the simple act of sending
a text message on your phone. This may feel like a task in which you engage in a goal-directed way, but its true habitual nature will be betrayed
by a change in circumstances, such as when you buy a new smartphone
or when you install a software update. First, you must unlock the phone—a
simple enough task until new software or hardware requires you to push
a different button or execute a different kind of swipe, in which case your
habitual actions frustratingly lead you to instead change the volume or
enter camera mode. Navigating to your messaging app can be similarly
problematic if the layout of your home screen changes after an update,
and you might find yourself inadvertently launching the wrong app. Your
fluid typing skills might be scuppered if the old button for punctuation
becomes the new button for emoticons. Even the way you interact with
people during a call is likely habitual. I still routinely answer my phone
as I always have, with a quizzical "Hello?"—despite the fact that my phone
now tells me exactly who is calling before I answer.

Even the way you use your hands to manipulate your phone is habitual. Although moving a finger might seem like a trivial act, it is in fact
quite a complex task. Because we have more muscles than we strictly
need, there are infinitely many ways we can utilize our muscles to achieve
a given movement. The exact pattern that we select seems to be carefully
chosen to be one that minimizes effort.[3] This finely tuned behavior turns
out to be habitual, however. In a remarkable experiment, Aymar de Rugy
and colleagues inflicted an injury to one of the wrist muscles of (consent-
ing!) human volunteers, weakening it relative to adjacent muscles.[4] The

appropriate, goal-directed response would be to use the injured muscle less and let noninjured muscles take over its share of the pulling. Instead, people used the same habitual coordination pattern as before, working the injured muscle even harder than before despite such work being more effortful and risking further injury.

Habitual and goal-directed behaviors have been shown to have distinct underlying neural bases. Through the kinds of navigation experiments in rats described above, it has been found that habitual behavior and goal-directed behavior depend on distinct underlying brain circuits.[5] Both types of behavior depend on an interaction between the cortex and the basal ganglia but on different subregions of these structures. The dorsomedial striatum, a subregion of the basal ganglia, seems to be critical to goal-directed behavior. If it is damaged, the rat's behavior will quickly become habitual even after limited exposure to a new task. Conversely, damage to the neighboring dorsolateral striatum will lead the rat to be more goal-directed and it will never lapse into habit, even after prolonged training. Similar behavioral changes can occur after damage to specific areas of the cortex. Analogous findings have been reported for monkeys learning to perform specific sequences of button pushes (much like typing a PIN)—different lesions can tip the balance of behavior toward being either habitual or goal-directed. Note that it doesn't really matter which exact parts of the brain are involved. What is important is that the balance *can* be shifted in different directions by inactivating different sites, demonstrating that goal-directed and habitual behavior each have dedicated regions within the brain devoted to them.

This finding suggests an intriguing, if perhaps drastic, solution to eliminating your unwanted habits. If you could inactivate just the right part of your dorsolateral striatum, you could be rid of them for good. Who wouldn't want to be goal-directed all the time?

Although it is true that the inflexibility of habits is a major drawback, this is outweighed by certain decisive advantages. Primary among these is their thoughtlessness. Once a maze can be navigated habitually, there is no need for the rat to expend cognitive resources trying to solve the maze. Instead, it is free to think and reason about other things or perhaps simply take a break from thinking. This advantage can be very important as behavior gets more complex. A hunter who no longer has to deliberate about how to launch his spear will be better able to track and anticipate the movements of his prey. Furthermore, the habit of throwing a spear can be deployed rapidly and automatically, without the need to wait for time-consuming thought processes. Since most circumstances in the real world are predictable and repetitive, the inflexibility of being habitual rarely matters. The risk of occasionally doing the wrong thing is a small price to pay for freedom of thought and rapidity of action.

There appears to be no limit to the number and complexities of habits we are capable of acquiring. Professional tennis players and soccer players must combine precise technique with awareness of other players' movements and make rapid, complex decisions about exactly what to do next. Such mastery can be achieved only if the basic elements of their skills (like correct technique) and even many of the more complex elements (like choosing the right pass or shot at the right time) have been forged into habits through years of practice. The habitual nature of the components of athletic skill is occasionally revealed when a top-flight player tries to modify his or her technique. Perfectionism drives many golfers to remodel their action in a quest for the ultimate swing. This process can take over a year and rarely ends well.[6]

Impressive as elite athletes may be, we should not underestimate that our own abilities to walk, talk, read, write, type, tie our shoelaces, do mental arithmetic, operate a phone, and play video games are extraordinary accomplishments in their own right. We all share a remarkable repertoire of learned behaviors, acquired through a lifetime of practice. As each new ability is learned and consolidated into habit, we move on to mastering the next one, building a vast library of skills that can be invoked in an instant, enabling our thoughts and actions, in all their complexity, to operate mostly on autopilot. Atop this mass conglomeration of habits sits a thin sliver of cognitive deliberation that steers only the highest-level decisions that we ever need to make. And without habits, it would be quickly overwhelmed.

NOTES

1. R. J. Dolan and P. Dayan, "Goals and Habits in the Brain," *Neuron* 80 (2013): 312–325.
2. J. A. Ouellette and W. Wood, "Habit and Intention in Everyday Life: The Multiple Processes by Which Past Behavior Predicts Future Behavior," *Psychological Bulletin* 124 (1998): 54–74.
3. A. H. Fagg, A. Shah, and A. G. Barto, "A Computational Model of Muscle

Recruitment for Wrist Movements," *Journal of Neurophysiology* 88 (2002): 3348–3358.

4. A. de Rugy, G. E. de Loeb, and T. J. Carroll, "Muscle Coordination Is Habitual Rather Than Optimal," *Journal of Neuroscience* 32 (2012): 7384–7391.

5. F. G. Ashby, B. O. Turner, and J. C. Horvitz, "Cortical and Basal Ganglia Contributions to Habit Learning and Automaticity," *Trends in Cognitive Sciences* 14 (2010): 208–215.

6. http://www.espn.com/golf/story/_/id/8865487/tiger-woods-reinvents-golf-swing-third-career-espn-magazine.

RELATING

Interpreting Information in Voice Requires Brain Circuits for Emotional Recognition and Expression

Darcy B. Kelley

ALEX THE PARROT was a particularly famous bird on several accounts.[1] Alex was smart: when presented with a tray full of different objects and asked to select the four-cornered blue one, he could grab the blue square with his beak on the first try. Alex clearly understood the question. "What matter, four-cornered blue?" researcher Irene Pepperberg, the lead scientist who studied Alex, would ask, and Alex would say, "Wood."[2] So Alex not only understood the human word "matter" (in the sense of substance), but he could also answer the question of what the blue square was made of: wood and not glass or metal or another material. When Alex died in 2007 at the age of thirty-one, his obituary ran in the *New York Times*.[3]

Beyond Alex's intelligence, though, there was this parrot's particular voice. Voice comprises the distinctive acoustic features that allow a listener to identify the speaker (child or adult? man or woman? who?). If you didn't know he was a bird and you had only the audio to go by, you would assume that Alex was a person.[4] The human quality of his voice was especially remarkable because—unlike humans (or even frogs)—birds don't use their larynx to create their communication sounds but instead have evolved a different vocal organ, the syrinx. The muscles of the bird's syrinx are activated by neurons in the hindbrain that, in humans, would control tongue movements.[5]

How can these complex sounds be produced and understood by a bird brain? Parrots are famous for vocal imitation, and the big talkers pick up on the vocabulary, and even the accent, of their owners. This ability to imitate sounds is a special kind of experience-dependent change in the brain called vocal learning and is surprisingly rare.[6] Only a very few other birds are known to be vocal learners. Zebra finches and canaries learn their songs, but for other finches, songs are innate and not learned, as is the cock-a-doodle-do of the crowing rooster. Long-term pairs of male and female zebra finches also learn each other's individual calls, and because of these abilities, zebra finches are the preferred species—the lab rat, as it were—for understanding vocal learning.[7]

In mammals, vocal learning is also rare, found only in bats, dolphins, humans, and elephants. That elephants can imitate (and thus learn) sounds was a big surprise. This ability was discovered at a refuge in Africa when the mysterious sound of a truck revving up each night was traced to one particular elephant called Malika.[8] If you didn't know this, you would assume that a person was gunning the engine—a much more probable sound than an elephant imitating a truck noise.

We don't know how animals with big, complex brains—like parrots, elephants, or dolphins—manage the feat of vocal learning. Song birds such as the zebra finch are the only animal for which we have this information. Song learning has been studied since the 1970s, and we now know that specific populations of neurons located in the forebrain of the zebra finch are dedicated to learned song production. The wiring diagram for these song neurons is very similar to the forebrain neural circuits that drive language learning in humans.[9]

Humans are great at learned vocal communication. As a species, we are adept at responding to words, figuring out the meaning of sentences, and framing and producing an appropriate vocal response. Where did language—the ability to communicate complex information through speech—come from? Ideally we could figure this out by determining when language arose in human evolution. But words leave no trace in the fossil record; we don't know whether Neanderthals talked or sang, and we are the only living human primate on the planet. Our closest non-human primate relative, the chimpanzee, doesn't use words to communicate, nor does teaching chimps how to use sign language[10] spark the kind of spontaneous, emergent language that developed, for example, in deaf children in Nicaragua.[11]

Vocal communication, however, is much more than speech. We also broadcast our sex, emotions, and even our intentions using our voices, and we recognize other individuals using their voices. One nonhuman primate—the baboon—has also been shown to recognize the voices of other individuals and to combine that information with remembered outcomes of recent social encounters to make decisions, a cognitive ability that could have served as a springboard for language evolution in humans.[12] I wonder, though, whether this ability is really primate-specific. Even zebra finches learn their mate's calls using voice.[13] I also suspect Alex the parrot learned which research assistant could be counted on for treats and would sidle up when he heard that particular voice.

So what is voice? Decoding social information from voice is analogous to determining which musical instrument is producing a specific note from its particular timbre. Vibrating a speaker cone 440 times a second (440 Hz) produces the note A. We distinguish an A produced by an oboe from the same note produced by a cello using timbre, subtle acoustic qualities of the sound that are different from pitch or loudness.[14] As for musical timbre, you can figure out whether your mom is mad at you from her voice (voice perception), and you can learn to convey your feelings to her by modulating your own voice (voice production).

For example, when Irene Pepperberg says, "Good boy, good parrot" in a video with Alex that I often use to teach, I can tell that she is functioning in a particular mode of vocal communication: affectionate instruction.[15] Dr. Pepperberg and her assistants use a specific kind of intonation—motherese—when they talk to Alex. Motherese is what we use when talking to babies or to our pets.[16] "Good boy, good parrot" in the video is motherese: the words are stretched out and pitch rises. We think that motherese boosts vocal learning in babies, in part because it addresses a particular challenge for every language: determining from speech sounds where one word ends and the next begins.[17] Children pick up languages easily, but by the teens, a new language becomes harder to learn for most of us.[18] Unlike French or Spanish, I never heard any Portuguese until I was an adult. Portuguese remains impenetrable to me because I cannot tell when one word ends and the next begins, the inherent challenge of spoken word segmentation.[19]

Motherese is not just for mothers: Dr. Pepperberg's male research assistant also uses this speaking mode with Alex. By listening to the video,

you can also tell that the lead scientist is a woman and the other researcher is a man, though you might find it hard to describe exactly *how*
you know (pitch is not enough). Voice carries some information that is
innate and other information that can be interpreted through experience.
Alan Alda narrates the Alex video I use for teaching. If you are familiar
with his acting or narration, Alda's voice would give him away, though
again you might find it hard to describe exactly which acoustic cues in
his voice provide the identification.

Just as some people are face blind and can identify objects with ease
but cannot distinguish one face from another,[20] a few rare folks are voice
deaf.[21] Face blindness is called *prosopagnosia,* and voice deafness is *phonagnosia.* In an amusing interview, a voice-deaf man explains that while he
could never tell to whom he was talking on the telephone, everyone else
seemed to know *his* voice.[22] He even generated a hypothesis: his voice
was so especially distinctive that it was a snap for others to identify him
on the phone. How does he cope? He fakes it, just going along in the
conversation until the person says something about his brothers or the
house, and he realizes that this strange voice at the other end of the line
belongs to his mother. Voice is essential for relating effectively to other
people. If you are somewhat deaf, even a hearing aid may not help you to
recognize your granddaughter's voice when she calls, creating an awk-
ward conversation. Current hearing aids are designed for words but need
to be redesigned for conveying the individual, intentional, and emotional
cues of voice.

A specific region of the brain is devoted to representing faces.[23] By
measuring the activity of neurons in these face patches, we can specify
the visual features that identify faces.[24] When these features are subtracted from face images, they are no longer effective in driving neural
activity. However, we do not know which human brain regions are specifically devoted to representing voices. If we had this information, we
could use neural responses to voices to specify the acoustic features that
are essential for recognizing voices and then redesign hearing aids to
amplify these sound features.

I propose that we begin by focusing on a key feature of voice: it conveys and we recognize information about the speaker's sex, emotional
state, and romantic interest. We share the ability to distinguish sex and
availability for mating with all vocal communicators. An evolutionarily

ancient part of vertebrate forebrain, known as the central nucleus of the amygdala (CeA), is the key to this aspect of vocal communication because it has access to the neural circuits of the hindbrain that express emotion.[25] The CeA could very well be a brain region that contributes to voice recognition.

In rhesus monkeys, many neurons in the CeA are activated both by sounds used in social interactions and the speaker's accompanying facial expressions.[26] Bats, like monkeys, use specific, lower-pitched calls to communicate in various social situations.[27] Sound information in social calls is carried by nerve activity from the hindbrain to the forebrain. Different forebrain CeA neurons *respond* to different calls.[28] The CeA also appears to *determine* the vocal response because electrically stimulating this region evokes social calls.[29] The CeA is sometimes called the autonomic amygdala because it drives hindbrain neural circuits that regulate automatic changes in breathing, heart rate, and blood pressure that accompany emotion and arousal.[30] This link between voice perception/production and emotion provides the key to understanding how voice is represented in the brain. Our own amygdala, for example, is galvanized by a baby's call, an especially painful acoustic experience for mothers.[31]

Another vertebrate, a frog known as *Xenopus,* is an especially powerful experimental model system for studying voice because sound is the only way it can communicate socially. When Africa and South America were still one supercontinent (Gondwanaland, about 150 million years ago) and its ancestors lived only on land as adults, *Xenopus* took entirely to underwater life and modified its larynx and the hindbrain circuits that generate songs to adapt.[32] *Xenopus* had to reconfigure the hindbrain circuits inherited from terrestrial ancestors to suppress breathing while calling and thus avoid drowning.[33] Hindbrain circuits are so critical for life that they have been very hard to study; one bad experimental move and it's all over. Luckily, the basic architecture of hindbrain circuits is common to all vertebrates, including frogs, birds, bats, monkeys, and humans. In *Xenopus,* the circuits can be studied more easily because its brain doesn't need oxygen from breathing. An isolated brain, placed in a bath of the neuromodulator serotonin, can be induced to sing—to produce nerve activity patterns that match actual song patterns.[34] We discovered that simulating the *Xenopus* CeA in an isolated brain also evokes singing because this forebrain region activates the hindbrain neural circuits that produce different vocal patterns.[35] In an actual frog, damage to the CeA produces socially inappropriate responses to calls in males (frog phonagnosia). Males normally respond to the calls of fertile females with an ardent answer call. However, calling males with CeA damage become mute when they hear the female's call, a response that only normally happens when the male hears another male rival. For *Xenopus,* a functioning CeA is necessary to distinguish male from female voices. These experiments in a species evolutionarily distant from us show that the ability of the brain to recognize and respond to social, emotion-provoking information in communication sounds relies on an ancient part of the brain, the autonomic amygdala, or CeA.

The elements of voice in a baby's cry are immediately recognizable; as for the frogs, our recognition is innate. To understand how we learn specific voices we can turn to birds that learn their mates' sounds. Male and female zebra finches form long-term pair bonds and recognize their mate's call.[36] In the nest, a pair of zebra finches will communicate softly about how they will share caring for the chicks that day.[37] If one partner's turn to care for the chicks is delayed, their nest duets and their turn-taking speeds up. The brain areas essential for learning more complicated communication sounds (courtship songs) in finches have been identified,[38] but we still don't know which regions are involved for the simpler calls used during mate-specific recognition and negotiations. I bet that the bird CeA also plays an essential role in recognizing a mate's voice. To test these ideas, we are working with colleagues at Columbia to record auditory responses in the bird CeA together with a collaborator at Southern Denmark University to record auditory responses in the *Xenopus* CeA. When we figure out which acoustic cues in calls drive CeA neurons, we can apply a very stringent test to make sure these cues are essential by deleting them from recordings to see if mate recognition is destroyed.

Some fascinating questions can be asked using animals such as *Xenopus* and zebra finches. Can we identify the specific sound features that tell a frog CeA neuron, "You are listening to a sexually receptive female"? How do auditory-driven neurons in the frog CeA access different hindbrain circuits for socially appropriate vocal responses? Do a chick's begging calls activate the parents' CeA in the same way that a human baby's cry triggers the human amygdala? Understanding how the brain decodes information carried by the voice using the amygdala should provide a new window into vocal communication and spark new ideas about how its most complex form—human language—evolved.

NOTES

1. I. M. Pepperberg, *The Alex Studies: Cognitive and Communicative Abilities of Grey Parrots* (Cambridge, MA: Harvard University Press, 2009).
2. http://alexfoundation.org/about/dr-irene-pepperberg/.
3. http://www.nytimes.com/2007/09/10/science/10cnd-parrot.html?_r=0.
4. https://www.youtube.com/watch?v=WGiARReTwBw.
5. R. A. Suthers and S. A. Zollinger, "Producing Song: The Vocal Apparatus," *Annals of the New York Academy of Sciences* 1016 (2004): 109–129.
6. S. Nowicki and W. A. Searcy, "The Evolution of Vocal Learning," *Current Opinion in Neurobiology* 28 (2014): 48–53.
7. E. C. Perez, J. E. Elie, I. C. Boucaud, T. Crouchet, C. O. Soulage, H. A. Soula, F. E. Theunissen, and C. Vignal, "Physiological Resonance between Mates through Calls as Possible Evidence of Empathic Processes in Songbirds," *Hormones and Behavior* 75 (2015): 130–141.
8. J. H. Poole et al., "Animal Behaviour: Elephants Are Capable of Vocal Learning," *Nature* 434 (2005): 455–456.
9. A. Calabrese and S. M. N. Woolley, "Coding Principles of the Canonical Cortical Microcircuit in the Avian Brain," *Proceedings of the National Academy of Sciences* 112 (2015): 3517–3522.
10. L. A. Petitto, "On the Biological Foundations of Human Language," in *The Signs of Language Revisited: An Anthology to Honor Ursula Bellugi and Edward Klima,* ed. K. Emmorey and H. Lane (Mahwah, NJ: Lawrence Erlbaum, 2000), pp. 447–471.
11. A. Senghas, S. Kita, and A. Özyürek, "Children Creating Core Properties of Language: Evidence from an Emerging Sign Language in Nicaragua," *Science* 305 (2004): 1779–1782. In the 1970s and 1980s, Nicaragua created two schools for deaf children in which they were taught lip reading and finger spelling rather than an established sign language. Within these communities, however, the children created their own sign language, which became more sophisticated as each class taught the next. The children have provided researchers with an extraordinary opportunity to investigate the birth of a new language.
12. R. M. Seyfarth and D. L. Cheney, "The Evolution of Language from Social Cognition," *Current Opinion in Neurobiology* 28 (2014): 5–9.
13. Perez et al., "Physiological Resonance between Mates through Calls as Possible Evidence of Empathic Processes in Songbirds."
14. T. M. Elliott, L. S. Hamilton, and F. E. Theunissen, "Acoustic Structure of the Five Perceptual Dimensions of Timbre in Orchestral Instrument Tones," *Journal of the Acoustical Society of America* 133 (2013): 389–404.

15. https://www.youtube.com/watch?v=WGiARReTwBw.

16. A. Fernald and P. Kuhl, "Acoustic Determinants of Infant Preference for Motherese Speech," *Infant Behavior and Development* 10 (1987): 279–293.

17. P. K. Kuhl, "Early Language Acquisition: Cracking the Speech Code," *Nature Reviews Neuroscience* 5 (2004): 831–843.

18. C. E. Snow and M. Hoefnagel-Höhle, "The Critical Period for Language Acquisition: Evidence from Second Language Learning," *Child Development* 49 (1978): 1114–1128.

19. E. D. Thiessen, E. A. Hill, and J. R. Saffran, "Infant-Directed Speech Facilitates Word Segmentation," *Infancy* 7 (2005): 53–71.

20. M. Behrmann and G. Avidan, "Congenital Prosopagnosia: Face-Blind from Birth," *Trends in Cognitive Sciences* 9 (2005): 180–187.

21. D. Van Lancker et al., "Phonagnosia: A Dissociation between Familiar and Unfamiliar Voices," *Cortex* 24 (1988): 195–209.

22. http://www.npr.org/templates/story/story.php?storyId=128412201.

23. D. Y. Tsao et al., "A Cortical Region Consisting Entirely of Face-Selective Cells," *Science* 311 (2006): 670–674.

24. D. Y. Tsao et al., "Faces and Objects in Macaque Cerebral Cortex," *Nature Neuroscience* 6 (2003): 989–995.

25. N. Moreno and A. González, "Evolution of the Amygdaloid Complex in Vertebrates, with Special Reference to the Anamnio-Amniotic Transition," *Journal of Anatomy* 211 (2007): 151–163.

26. K. Kuraoka and K. Nakamura, "Responses of Single Neurons in Monkey Amygdala to Facial and Vocal Emotions," *Journal of Neurophysiology* 97 (2007): 1379–1387.

27. M. A. Gadziola, J. M. S. Grimsley, S. J. Shanbhag, and J. J. Wenstup, "A Novel Coding Mechanism for Social Vocalizations in the Lateral Amygdala," *Journal of Neurophysiology* 107 (2012): 1047–1057.

28. M. J. Clement et al., "Audiovocal Communication and Social Behavior in Mustached Bats," *Behavior and Neurodynamics for Auditory Communication* (2006): 57–84; D. C. Peterson, and J. J. Wenstrup, "Selectivity and Persistent Firing Responses to Social Vocalizations in the Basolateral Amygdala," *Neuroscience* 217 (2012): 154–171.

29. M. Jie and J. S. Kanwal, "Stimulation of the Basal and Central Amygdala in the Mustached Bat Triggers Echolocation and Agonistic Vocalizations within Multimodal Output," *Frontiers in Physiology* 5 (2014).

30. Ibid.

31. E. D. Musser, H. Kaiser-Laurent, and J. C. Ablow, "The Neural Correlates of Maternal Sensitivity: An fMRI Study," *Developmental Cognitive Neuroscience* 2 (2012): 428–436.

32. B. L. S. Furman et al., "Pan-African Phylogeography of a Model Organism, the African Clawed Frog 'Xenopus laevis,'" *Molecular Ecology* 24 (2015): 909–925; M. Tobias and D. B. Kelley, "Vocalizations of a Sexually Dimorphic

Isolated Larynx: Peripheral Constraints on Behavioral Expression," *Journal of Neuroscience* 7 (1987): 3191–3197.

33. E. Zornik and D. B. Kelley, "Regulation of Respiratory and Vocal Motor Pools in the Isolated Brain of *Xenopus laevis*," *Journal of Neuroscience* 28 (2008): 612–621.

34. H. J. Rhodes, H. J. Yu, and A. Yamaguchi, "Xenopus Vocalizations Are Controlled by a Sexually Differentiated Hindbrain Central Pattern Generator," *Journal of Neuroscience* 27 (2007): 1485–1497.

35. I. C. Hall, I. H. Ballagh, and D. B. Kelley, "The Xenopus Amygdala Mediates Socially Appropriate Vocal Communication Signals," *Journal of Neuroscience* 33 (2013): 14534–14548.

36. Perez at al., "Physiological Resonance between Mates through Calls as Possible Evidence of Empathic Processes in Songbirds."

37. I. C. A. Boucaud, M. M. Mariette, A. S. Villain, and C. Vignal, "Vocal Negotiation over Parental Care? Acoustic Communication at the Nest Predicts Partners' Incubation Share," *Biological Journal of the Linnean Society* 117 (2016): 322–336.

38. M. S. Brainard and A. J. Doupe, "What Songbirds Teach Us about Learning," *Nature* 417 (2002): 351–358.

Mind Reading Emerged at Least Twice in the Course of Evolution

Gül Dölen

WHEN WE THINK OF MIND READING, most people imagine the abilities of a magician, an evil genius, or an alien life form. In fact, we perform mind reading everyday. For example, as you are reading this book, maybe you are on the subway, and there are no open seats, so you are standing. The train pitches and sways, and from the corner of your eye you notice that a nearby passenger starts to gather up his belongings. You check around for other passengers who might also be eyeing the seat and note that you are the only one who seems to have noticed. So without drawing too much attention to yourself, you move in closer. When the passenger gets up, you quickly swoop in and grab the seat. You can now continue reading this book in comfort.

In carrying out this simple strategic move for a seat, you have performed several acts of mind reading. Neuroscientists call mind reading of this sort "Theory of Mind" (abbreviated ToM, also sometimes called "intentional stance," "social cognition," "cognitive empathy," "mind attribution," and "mentalizing"). ToM is a cognitive ability that is only a theory insofar as it is a best guess about another person's mental states (that is, desires, beliefs, knowledge) and an understanding that these mental states might be different from your own. As in the example above, we use this theory to understand other people's motivations, anticipate their actions, and modify our own behaviors accordingly. We perform ToM mental tasks constantly and often reflexively, without conscious awareness.

Consider, for example, the Heider-Simmel illusion.[1] An animated film shows a large rectangle, a big triangle, a small triangle, and a small circle moving around a page. When asked to describe what the film is about, most people will ascribe thoughts, feelings, and emotions to the two triangles and the circle and construct a scenario explaining interactions among them—for example, the small triangle is annoying, the big triangle is a bully, the circle is a dreamer.

Although most of us perform mind attributions easily and reflexively, as with many of our cognitive abilities, we are not born with ToM. Rather, children begin to demonstrate such abilities starting around age four. As children incorporate ToM into their play behavior, they will typically attribute desires, beliefs, and knowledge to dolls and figurines. As ToM abilities become increasingly more sophisticated, children begin to use their newly acquired abilities for joking and pranking. If you spend any time with children, you may have been the victim of the classic shoulder tap prank (where the child reaches around your back and taps you on the opposite shoulder, causing you to look the wrong way); the prank requires that the child understand that your knowledge of the source of the tap is different from the child's own.

Most of what we know about the brain mechanisms that enable ToM comes from studies in humans. Brain imaging studies of human subjects have consistently correlated performance of ToM tasks with activity in two brain regions: the right temporoparietal junction and the dorsomedial prefrontal cortex.[2] It is not surprising that these brain regions are also activated when viewing the Heider-Simmel illusion described above.[3]

Many scientists have held that ToM abilities are restricted to humans.[4] Anyone with pets, especially dogs, would be incredulous at this assertion because so much of the behavior exhibited by these companion animals seems to meet the criteria for having a ToM. There are three theoretical problems for scientists wanting to formally claim that animals have this attribute.

First, as we've already discussed, reflexive mind attribution is difficult to turn off, even when we know the objects (such as circles, triangles, dolls, or figurines) are incapable of desires, beliefs, or knowledge. Thus when objects are not objects at all, but rather living, breathing, companions, it is hard to know whether we see feelings, emotions, and thoughts in our pets because they really have them or because our brains are wired to believe that they have them.

The second problem for scientists who want to argue that animals have a ToM is how to demonstrate it to a skeptic. For example, we might be tempted to say that a dog brings you his leash when he wants to be taken out because he anticipates that your inability to find it will delay departure and that this is an effective way to communicate to you his desire to go out. On the other hand, a skeptic might argue that this behavior is just an elaborate sequence of actions that the dog has learned to produce the desired outcome and that you are just a fancy kind of robot that he must include in the sequence.

The third problem for animal ToM is that in terms of brain anatomy, only humans are thought to be equipped with the required brain regions (the aforementioned temporoparietal junction and the dorsomedial prefrontal cortex) implicated in ToM. Although studies in monkeys have revealed that another region, called the superior temporal sulcus, may correspond to the temporoparietal junction in humans,[5] the dorsomedial prefrontal cortex is proportionally much larger in humans, even when compared to our closest primate relatives.[6] This is because during the course of evolution, the human brain, and especially the cerebral cortex (the folded outer surface that contains both the temporoparietal junction and the dorsomedial prefrontal cortex), underwent a massive expansion, both in relative size and in number of neurons.[7]

So how might we begin to tackle the scientific problem of whether animals have ToM? Let us begin with how it is demonstrated in humans. The vast majority of studies use some variant of the Sally-Ann False Belief task. In this instance, the subject is presented with a vignette consisting of two characters, Sally and Ann. Sally has a ball and places it into her basket and then leaves the room. Ann then removes the ball from Sally's basket and places it into her own. Then Sally returns, and the subject is asked a series of questions: (1) Who is Ann? (2) Who is Sally? (3) Where is the ball? (4) Where does Sally think the ball is? A typically developing child, starting at around age four, will correctly answer all four questions.

When tested on the Sally-Ann False Belief task, children with autism spectrum disorder will give the same answer for both questions (3) and

(4) because these children are unable to distinguish between their own and Sally's knowledge of the location of the ball. These ToM impairments in autism spectrum disorder are not due to diminished intelligence because they are evident even in high-functioning autistics, and conversely, Down's syndrome patients, even those with severe intellectual disability, show no ToM deficits.[8]

The cause of autism is overwhelmingly genetic. To date, mutations in over 880 genes have been associated with causing autism, and future studies will likely reveal many more. While it is still unknown how each of these gene disruptions produces the symptoms of autism, most of these genes are also found in animals. Since these genes are required for ToM abilities in humans (that is, mutations lead to ToM deficits in autism), they may also be important for ToM-like functions in other animals. If so, we would say that the genes found in humans and other animals are homologues because they encode the same function (ToM) across evolutionarily related species.

In fact, it is unlikely that ToM abilities in humans emerged spontaneously. Instead, they probably evolved from existing cognitive abilities. Researchers agree that many of the foundational capacities of ToM are also exhibited by a number of other animals, including dolphins, some primates, and scrub jays. For example, dolphins recognize themselves in mirrors and are able to differentiate between individual members of their social group. Because the ability to attribute mindfulness to others requires being able to differentiate between self and other, these abilities are said to be foundational capacities of ToM.

We humans mainly use ToM in a social context; therefore most researchers believe that this cognitive function evolved in response to the selection pressures imposed by social living.[9] By this reasoning, ToM-like abilities might only be observed in species that live socially (for example, humans, most primates, dolphins, and wolves).

Nevertheless, we can imagine that the ability of a predator to catch its prey would be greatly enhanced if the predator were to attribute mindfulness to the prey. This idea raises the interesting possibility that *predatory* rather than *social* selection pressures could have shaped the evolution of ToM. If so, ToM-like abilities would be much older (in terms of evolutionary time) than has previously been supposed since hunting almost certainly evolved earlier than social living.

Supporting this latter view, the recently rediscovered Larger Pacific Striped Octopus exhibits a unique hunting behavior that suggests that these animals might have ToM-like cognitive abilities.[10] When hunting shrimp, this octopus uses what appears to be a variant of the shoulder-tap prank. Having identified its prey, the octopus assumes a dark-colored skin pattern and then very slowly extends one of its dorsal arms, arching it above and around the shrimp. When the octopus is in position, it lowers the tip of its extended dorsal arm behind the shrimp and taps it on its pleomere (the posterior portion of the shrimp's body), causing the shrimp to leap forward into the octopus's other seven arms.[11]

The Larger Pacific Striped Octopus is one of a handful of known *social* species of octopus (the vast majority of the three hundred or so members of the order *Octopoda* are solitary and cannibalistic). Nevertheless, in the example described above, this octopus's ToM-like behavior is used in a predatory setting. Furthermore, the ToM-like hunting strategy is also used by this octopus's *asocial* sister species, *Octopus chierchiae*. Anthropomorphic descriptions notwithstanding, the hunting strategy used by *Octopus chierchiae* to catch shrimp is reminiscent of the ruse of the man at the movies, yawning and stretching in order to get an arm around his date. Taken together, these observations support the hypothesis that, at least in octopuses (yes, octopuses, not octopi), predatory rather than social selection pressures molded the emergence of ToM-like behaviors.

Although the similarity of the Larger Pacific Striped Octopus's hunting strategy to the shoulder-tap prank may merely represent an especially convincing Heider-Simmel illusion, this animal is able to modify its strategy in a context-specific way, suggesting that it has true ToM. For example, if the prey's view is obstructed (for example, when the octopus is hunting hermit crabs, whose shells cover them from behind), this octopus uses a direct pounce strategy rather than the pleomere tap strategy, suggesting that it knows the difference between a prey that can see its predator and one that does not, and it acts accordingly. Although these observations may not convince a skeptic, they do suggest that the octopus is employing a flexible cognitive strategy rather than just an elaborate set of learned routines.

Still the problem of brain anatomy remains: the octopus brain is nothing like a human brain. It doesn't have any of the cortical regions suggested to underlie ToM or even, for that matter, a cerebral cortex. Nevertheless, the degree of encephalization and neuronal expansion seen in octopuses sets this species apart from other invertebrates. Indeed, the nervous system of the octopus, thought to be the most intelligent invertebrate, is comprised of approximately half a billion neurons, more than six times the number in a mouse brain. Additionally, like humans, dolphins, and elephants, octopuses have a brain with a folded surface, ostensibly to pack in more neurons in a confined space, in contrast to the smooth-surfaced brains of other cephalopods, mice, rats, and marmosets. Thus although octopuses don't have cortical regions associated with ToM, they have an exceptionally large brain capacity and may have evolved to solve the problem of ToM using different anatomical strategies.

Perhaps the most effective way to determine whether octopuses have ToM would be to show that the same genes that are necessary for this attribute in humans are also necessary for ToM-like behaviors in octopuses. Experiments to test this possibility are under way in my lab. In the meantime, it is interesting to note that octopuses have gene homologues for at least two of the genes most centrally implicated in autism.[12] Although it may seem far-fetched to suppose that these genes may encode ToM in octopuses, there is in fact precedence for this type of so-called deep homology. For example, the invertebrate compound eye is anatomically very different from the vertebrate camera eye, and vertebrate and invertebrate lineages diverged from each other long before the evolution of either type of eye. Nevertheless, the *Pax6* gene, found in vertebrates and invertebrates, is required for the formation of both types of eye. Most recently, a set of genes has been described that controls language in both humans and African grey parrots, despite anatomical differences in brain organization between humans and birds and the absence of a common ancestor that shares the language trait.[13] If manipulation of genes implicated in autism impairs ToM-like hunting behaviors in octopuses, such a finding would provide a novel example of deep homology. Moreover, because it would mean that octopuses are able to use the same genes as humans to encode ToM-like behaviors, even though their brains are organized totally differently, such a finding would suggest that it is the genes, rather than the specific brain circuit arrangements, that confer complex brain function across species.

NOTES

1. F. Heider and M. Simmel, "An Experimental Study of Apparent Behavior," *American Journal of Psychology* 57 (1944): 243–259. To see how you would respond to the illusion, try it out for yourself on one of the videos of the Heider-Simmel illusion available on the Internet, such as https://www.youtube.com/watch?v=8FIEZXMUM2I.

2. R. Saxe and N. Kanwisher, "People Thinking about Thinking People: The Role of the Temporo-Parietal Junction in 'Theory of Mind,'" *Neuroimage* 19 (2003): 1835–1842; C. D. Frith and U. Frith, "Interacting Minds—A Biological Basis," *Science* 286 (1999): 1692–1695.

3. F. Castelli, C. Frith, F. Happé, and U. Frith, "Autism, Asperger Syndrome and Brain Mechanisms for the Attribution of Mental States to Animated Shapes," *Brain* 125 (2002): 1839–1849.

4. R. Saxe, "Uniquely Human Social Cognition," *Current Opinion in Neurobiology* 16 (2006): 235–239.

5. R. B. Mars, J. Sallet, F.-X. Neubert, and M. F. S. Rushworth, "Connectivity Profiles Reveal the Relationship between Brain Areas for Social Cognition in Human and Monkey Temporoparietal Cortex," *Proceedings of the National Academy of Sciences of the USA* 110 (2013): 10806–10811.

6. H. B. Uylings and C. G. van Eden, "Qualitative and Quantitative Comparison of the Prefrontal Cortex in Rat and in Primates, Including Humans," *Progress in Brain Research* 85 (1990): 31–62.

7. F. A. C. Azevedo et al., "Equal Numbers of Neuronal and Nonneuronal Cells Make the Human Brain an Isometrically Scaled-Up Primate Brain," *Journal of Comparative Neurology* 513 (2009): 532–541.

8. S. Baron-Cohen, A. M. Leslie, and U. T. A. Frith, "Does the Autistic Child Have a 'Theory of Mind?'" *Cognition* 21 (1985): 37–46.

9. J. Decety and M. Svetlova, "Putting Together Phylogenetic and Ontogenetic Perspectives on Empathy," *Developmental Cognitive Neuroscience* 2 (2012): 1–24.

10. A. F. Rodaniche, "Iteroparity in the Lesser Pacific Striped Octopus *Octopus chierchiae*," *Bulletin of Marine Science* 35 (1984): 99–104; R. L. Caldwell, R. Ross, A. Rodaniche, and C. L. Huffard, "Behavior and Body Patterns of the Larger Pacific Striped Octopus," *PLoS One* 10 (2015): 1–17.

11. Caldwell, Ross, Rodaniche, and Huffard, "Behavior and Body Patterns of the Larger Pacific Striped Octopus."

12. C. B. Albertin et al., "The Octopus Genome and the Evolution of Cephalopod Neural and Morphological Novelties," *Nature* 524 (2015): 220–224.

13. A. R. Pfenning et al., "Convergent Transcriptional Specializations in the Brains of Humans and Song-Learning Birds," *Science* 346 (2014): 1256846.

We Are Born to Help Others

Peggy Mason

I LOVE DISASTER MOVIES. Anyone who has ever watched *Airplane* (1970), *The Poseidon Adventure* (1972), *Armageddon* (1998), *White House Down* (2013), or their ilk knows that the true draw of these thrilling action movies with formulaic scripts is the underlying human drama.[1] We're introduced to people whom we come to care about, knowing full well that not all will survive. But those who die will go out in a blazing magnificence of self-sacrifice to save others. We feel good on watching people helping others, despite the inevitably awful outcomes. Special effects may be the shiny attraction, but the irresistible pull of disaster movies is the opportunity to vicariously experience the joy of one individual helping another. Whereas noble screen heroes appear to embody the essence of being human, I argue here that they are in fact emblematic of an evolutionary ideal common to the hairy beasts on this planet, the mammals.

The act of helping is a fundamental mammalian behavior. Mammals are a group of animals identified by characteristics including their having hair and depending on mother's milk at birth and for a period of time thereafter.[2] Newborn mammals are incapable of surviving alone. They depend absolutely, unequivocally, on mom. Without a mother, newborns cannot survive without modern interventions. Because mammals depend on mother's milk, all mammals need help at the start of their lives.

A mother's care for her offspring extends beyond simply providing nutrition.[3] As they nurse, mothers keep their offspring warm. This is necessary since newborn mammals don't move around much, leaving them with a limited capacity to produce sufficient metabolic heat for growth and development. Mammalian parents, typically mothers, provide shelter that ensures their offspring's safety from predators and adverse weather. In most mammals, mother's milk is also essential to provide a starter immune system, without which newborns fall victim to fatal infectious diseases at a high rate. While humans and several other primates transfer antibodies during pregnancy rather than postnatally, there remains an overwhelming evolutionary advantage to being successfully nursed.

The upshot of nursing and the associated mothering is that well-nursed offspring grow up into healthy adults, and poorly nursed offspring either die or fail to thrive.[4] Throughout evolutionary time, this dichotomy resulted in more offspring born to the well nursed than to the poorly nursed. Thus caring maternal behavior became *de rigueur* among mammals from echidnas to koala bears, pandas, weasels, and, of course, humans and other primates.

Despite the comprehensive care provided by mothers, a warm, immune-boosting, milk-providing protector is not enough to ensure that a baby will grow into a normal, socially adjusted adult. In his decades of work with rhesus macaque monkeys, the American psychologist Harry Harlow showed that social interactions, particularly with peers, were critical to normal social development.[5] Monkeys raised in a warm cage with unlimited access to milk and complete protection from potential dangers —conditions that one may consider ideal for primate physiology—became emotional cripples if deprived of the opportunity to play with other young monkeys.

Monkeys raised from birth in total social isolation became socially weird.[6] They were either fearless or suffered from crippling fear; none walked that middle ground of showing fear only when appropriate. As adults, they retreated from normally raised monkeys. Socially isolated monkeys displayed repetitive behaviors reminiscent of autism, such as rocking or self-hugging. Some refused to eat, even to the point of requiring a feeding tube to avoid death. They failed to mate, and, when impregnated through artificial insemination, females did not care for their offspring.

The take-home message from Harlow's experiments is that it takes more than healthy lungs, heart, or kidneys to succeed at being a mammal who flourishes within the environment. Mammalhood is not simply a physical matter, but also requires the effects of socialization on brain circuits. Furthermore, it is not only future mothers who are being socialized. All offspring, both males and females, learn to get along with others. This foretells the general importance of sociality. Adult mammals accrue advantages from group living because they are better able to find food. Group living also increases the chances that individuals will have adequate shelter or roost and remain safe from predators. One of the glues to group living is pro-social behavior or helping.

The "Christmas card" experiment of sociologists Phillip Kunz and Michael Woolcott dramatically demonstrates the power of receiving to motivate giving.[7] In this study, 578 names were chosen randomly out of a printed directory. Christmas cards signed by the first author and his wife were sent to these complete strangers. It is remarkable that more than a hundred recipients (about 20 percent) sent back cards. Some sent detailed warm notes, complete with photographs, to the Kunz family. In an interview on National Public Radio, Dr. Kunz revealed that he continued to get cards from some of the random recipients for fifteen years running.[8] The reciprocity effect is so strong for humans that charities use it to increase donations. Donation rate reliably increases by as much as 35 percent when a charity sends a gift (such as address labels) along with a solicitation for money.[9]

Reciprocal helping also exists in other animals. One example is female vampire bats, who share blood with other female vampire bats. Within a stable group of captive bats, universally familiar with each other but of mixed genetic relatedness, the greatest predictor for blood sharing was whether or not a donor bat had previously received blood from a particular bat.[10] In other words, bats preferentially gave blood to bats who had given them blood previously. This reciprocity was not generalized to the entire bat community. Receiving blood did not make a bat more likely to provide blood to just any other bat; it specifically increased the likelihood that she would provide blood to the same individual who had given her blood.

Reciprocal helping only goes so far and fails to explain help that is offered during times of disaster. In the face of people endangered by a hurricane, earthquake, or similar catastrophe, people often provide help at great potential personal cost. Nonetheless, people even help total strangers in these situations. Why does this happen?

If reciprocity, kinship, and self-interest are not motivating factors, what could possibly lead people to help? One potential answer is that people help because they think through the issues and potential consequences of possible actions, ultimately concluding that helping is the right and moral course of action. If this were the case, then, clearly, helping would be limited to humans. However, anecdotes abound and experimental evidence confirms that apes and monkeys also help by providing food, tools, aid, and comfort to other apes and monkeys.[11]

The fact that nonhuman primates aid unrelated others firmly disputes the idea that helping requires cultural transmission through religious, moral, or educational institutions. Indeed, in a study of more than one thousand primarily Muslim and Christian children from countries around the world, religiosity was *inversely* related to children's generosity in sharing stickers.[12] Children from more religious homes gave away fewer stickers than did children from less religious backgrounds. Of great interest is that religiosity was *positively* correlated with the children's interest in punishing perceived bad behavior.

In the past ten years, it has become clear that the roots of helping extend further into the mammalian tree, well beyond primates. Indeed, rodents, which diverged from the primate line more than seventy million years ago, show pro-social behavior. For example, rats help other rats escape from unpleasant places even if the trapped rats are complete strangers.[13] When given a choice between delivering food to another rat in addition to themselves or providing only for themselves, rats choose the former.[14] Despite their enormous differences, humans and rats share the capacity to detect and respond helpfully to the distress of others.

You may be thinking, "*This is all well and good, but I am sure that humans are way better at helping than rats or even chimps.*" As it turns out, chimps and young children (less than four years of age) both help strangers (either other chimps or adult humans) at comparable rates.[15] In a fascinating twist (that would make Machiavelli smile) children do not become more helpful over time but rather become more discriminating about whom they are willing to help. The same is true of chimps. With experience, chimps and children become more likely to help others who have helped them, a nod to reciprocity. It is interesting that only children become more selective regarding helping members of other social groups. In essence, children learn not to help certain categories of others.

Martin Luther King said, "Life's most persistent and urgent question is: What are you doing for others?" The answer is written in our biological inheritance. We, along with our mammalian cousins, are drawn to helping others in distress.

NOTES

1. It is not lost on the author that women are absent as heroes of disaster movies. This is particularly ironic since most females are thought to be slightly more empathic than most males, as explored in the following review: L. Christov-Moore, E. A. Simpson, G. Coudé, K. Grigaityte, M. Iacoboni, and P. F. Ferrari, "Empathy: Gender Effects in Brain and Behavior," *Neurosciences and Biobehavioral Reviews* 46 (2014): 604–627.

2. The time when mammalian offspring are dependent on mother's milk is termed the weaning period. The length of this period is different across species. In rats and mice, it is roughly 3–4 weeks. Chimps nurse until they are roughly four years old, about eight months after their first molars appear. In human children, the first teeth (typically not molars) appear at about 4–6 months. When human weaning would occur under natural, premodern conditions is not definitively known; the earliest estimates are 6–9 months and the latest are a few years; T. M. Smith, Z. Machanda, A. B. Bernard, R. M. Donovan, A. M. Papakyrikos, M. N. Muller, and R. Wrangham, "First Molar Eruption, Weaning, and Life History in Living Wild Chimpanzees," *Proceedings of the National Academy of Sciences of the USA* 110 (2013): 2787–2791.

3. J. K. Rilling and L. J. Young, "The Biology of Mammalian Parenting and Its Effect on Offspring Social Development," *Science* 345 (2014): 771–776.

4. These days, providing milk and care to a newborn are divisible tasks that can be accomplished by individuals other than the mother. Thus fathers can feed a baby and mothers are freed from the absolute obligation of providing nutrition to newborns at all times during the nursing period. Nonetheless, there is a sensory "conversation" that takes place between a nursing mother and suckling newborn. For example, as milk is consumed, supplies run down, and the baby must suck harder to stimulate more milk let-down. This initial interaction between mother and baby is thought to set the stage for adult social and eating behavior. Indeed, bottle-fed babies consistently weigh more than do breast-fed babies; C. P. Cramer and E. M. Blass, "Mechanisms of Control of Milk Intake in Suckling Rats," *American Journal of Physiology* 245 (1983): R154–159.

5. M. Vicedo, "Mothers, Machines, and Morals: Harry Harlow's Work on Primate Love from Lab to Legend," *Journal of the History of the Behavioral Sciences* 45 (2009): 193–218.

6. H. F. Harlow, R. O. Dodsworth, and M. K. Harlow, "Total Social Isolation in

Monkeys," *Proceedings of the National Academy of Sciences of the USA* 54 (1965): 90–97.

7. P. R. Kunz and M. Woolcott, "Season's Greetings: From My Status to Yours," *Social Science Research* 5 (1976): 269–278.

8. A. Spiegel, "Give and Take: How the Rule of Reciprocation Binds Us," http://www.npr.org/sections/health-shots/2012/11/26/165570502/give-and-take-how-the-rule-of-reciprocation-binds-us.

9. Robert Cialdini studied Hare Krishna devotees who gave people a small token, such as a flower, and then asked for a donation. The rate of donations increased tremendously (over rates without such a token gift), although many token recipients were visibly angry. While they presumably felt tricked into *needing to reciprocate,* Cialdini observed that they nonetheless did in fact reciprocate. This speaks to the strength of the rule of reciprocity; R. B. Cialdini, *Influence: The Psychology of Modern Persuasion* (New York: Harper Trade, 1984).

10. C. G. Carter and G. S. Wilkinson, "Food Sharing in Vampire Bats: Reciprocal Help Predicts Donations More Than Relatedness or Harassment," *Proceedings of the Royal Society Series B* 280 (2013): 20122573.

11. Frans de Waal relates many examples of empathic behavior in nonhuman primates. The stories are remarkable and deeply reminiscent of behavior of humans at their best. Lest you think that these stories are too anthropomorphic to warrant serious consideration, remember that rejecting the evolutionary connection between apes and humans (anthropophobia?) is as dangerous as overplaying it; F. de Waal: *Peacemaking among Primates* (Cambridge, MA: Harvard University Press, 1989), and *The Age of Empathy* (New York: Three Rivers Press, 2010).

12. In this experiment, children received thirty stickers, from which they chose their ten favorites. The children were then told that the experimenters had run out of time and not all the children would receive stickers. They could then give away stickers to other children, anonymous children from the same school and cultural and religious background. Children from more religious homes gave away fewer stickers. It is interesting that the most religious parents assessed their own children as more generous than did the nonreligious parents. In fact, the opposite pattern was observed; J. Decety, J. M. Cowell, K. Lee, R. Mahasneh, S. Malcolm-Smith, B. Selcuk, and K. Zhou, "The Negative Association between Religiousness and Children's Altruism across the World," *Current Biology* 25 (2015): 2951–2955.

13. In the protocol introduced by Inbal Ben-Ami Bartal and colleagues, a rat was placed in a Plexiglas tube that was just big enough to allow the trapped rat to turn around. However, the door to the tube could be opened only from the outside. Roughly 75 percent of rats outside the tube who were given the opportunity to open the door to the tube did so and thereby liberated the trapped rat. This rate of helping did not depend on individual familiarity; rats helped known cage mates and strangers alike; I. Ben-Ami Bartal, J. Decety, and

P. Mason, "Empathy and Pro-Social Behavior in Rats, *Science* 334 (2011): 1427–
1430; I. Ben-Ami Bartal, D. A. Rodgers, M. S. Bernardez Sarria, J. Decety, and
P. Mason, "Pro-Social Behavior in Rats Is Modulated by Social Experience,"
Elife 3 (2014): e01385. In the protocol used by Nobuya Sato and colleagues,
one rat was placed in a pool of water. A door from the pool to a dry compart-
ment could be opened only by a free rat located on the dry side of the door.
Free rats opened for a "soaked" rat but did not open when the pool was dry.
This shows that the rats opened only for rats in distress and did not open the
door simply to play with another rat; N. Sato, L. Tan, K. Tate, and M. Okada,
"Rats Demonstrate Helping Behavior toward a Soaked Conspecific," *Animal
Cognition* 18 (2015): 1039–1047.

14. Two independent laboratories used similar methods to examine whether rats
would choose to provide food to another rat. Happily, the two laboratories had
nearly identical findings and published their reports within months of each
other; J. Hernandez-Lallement, M. van Wingerden, C. Marx, M. Srejic, and
T. Kalenscher, "Rats Prefer Mutual Rewards in a Prosocial Choice Task,"
Frontiers in Neuroscience 8 (2015): 443; C. Márquez, S. M. Rennie, D. F. Costa,
and M. A. Moita, "Prosocial Choice in Rats Depends on Food-Seeking
Behavior Displayed by Recipients," *Current Biology* 25 (2015): 1736–1745.

15. Of course, chimps help only on tasks that are understandable to them, such
as reaching a desired object that is out of reach or unlatching a door. They
don't help fix a toy. Chimps also do not divide up and share food, although
they will pull a lever that provides themselves and another with separate
pieces of food; F. Warneken and T. Tomasello, "Varieties of Altruism in Children and Chimpanzees," *Trends in Cognitive Sciences* 13 (2009): 397–402.

Intense Romantic Love Uses Subconscious Survival Circuits in the Brain

Lucy L. Brown

"WHAT IS LOVE?" In a single month of 2016, four hundred thousand people around the world sat at their computers and asked Google that question. *Psychology Today* was at the top of the search results page with this answer: "Love is a force of nature. However much we may want to, we cannot command, demand, or take away love, any more than we can command the moon."[1]

Love is indeed a force of nature. In neuroscientific terms, intense romantic love engages brain circuits that operate at a subconscious level. These neural circuits are so critical that they are near areas in the brain stem that mediate our most basic survival reflexes such as breathing and swallowing. This location for love's essential systems helps to explain why we feel so out of control in the early stages of romantic love or after losing a lover. We have little control over our feelings and actions about love, just as we feel compelled to locate and swallow water when we are thirsty.

We may say "I love you" to people we are dating, spouses, parents, our children, siblings, and friends—even pets. The kind of love I am addressing here, the involuntary force that can so easily upend our lives, is the early-stage, intense romantic love that craves emotional union with another person, even more than sex. It is the "How do I love, thee? Let me count the ways" type of love—the consuming love that makes us think about the other person for most of our day.[2] The psychologist Dorothy Tennov called it "Limerance."[3] Elaine Hatfield developed a questionnaire to measure it, and she called it the Passionate Love Scale.[4] Other psychologists looked at Western literature and developed a Love Attitudes Scale that has six different types of love, from Eros (obsessive passion) to Storge (friendship-based love).[5] These are all aspects of romantic love, and measuring them is useful to show both the variations and predictable nature of romantic love and to help organize the search for the brain systems involved.

Helen Fisher, an anthropologist, has placed early-stage romantic love in an evolutionary context. She sees it as part of the evolved human strategy to pass on your genes: have sex, have offspring, and then protect yourself and your offspring as they grow up.[6] This view is helpful for understanding romance and argues against those who think romantic love is nonexistent, magical, or unimportant. Fisher's proposal is that there are three essentials for human reproduction: lust; attraction/romance; and attachment. "Lust" is the sex drive, a primitive system that we all accept as necessary for survival of the species. "Attraction" is a behavior that we can observe in other animals as they court, like female peahens as they approach the male peacock and his magnificent tail. This courtship behavior and attraction response is just as essential as the reproductive sexual act to the survival of their species. The important thing about attraction in humans is the focus on one person over months, kept alive by the euphoria that we call romance, and the protection from harm that another person can provide. It is a mate choice, which through our evolution has kept a pair together long enough to get a child to be independent from the mother for nourishment.[7] Long-term "attachment" contributes to the success of the family group, which is advantageous for the further protection of individuals.

In this context, several of us, including Helen Fisher, decided to investigate the neurobiology of early-stage, intense romantic love to learn why people change their lives to be together and because psychologists disagreed about love's definition and origins—that is, is it an emotion or a drive?[8] An emotion is a strong feeling, and romance is certainly that. The emotion of euphoria associated with romance can come and go, however, and be replaced by other emotions like anxiety or anger (jealousy), so it might be inappropriate to identify romance as a single emotion. A drive is longer lasting than emotions; also, drives have goals, like eating food and drinking water, and the obsession with another person makes romance look more like a drive: being with the other person is the goal. Neuroscience has identified different brain regions to be associated with emotions and goals, so brain-imaging studies might help to classify romance.

The major finding of our first brain-scanning study was that everyone who was in the early stages of romance and had a high score on the Passionate Love Scale activated a key brain stem reward-and-drive region when he or she looked at a picture of the beloved's face.[9] This dopamine-using region is called the ventral tegmental area. We need this region to know that water relieves the unpleasant feeling of thirst or that food relieves the unpleasant feeling of hunger. We cannot live without this system. Other parts of the brain related to reward were activated too, such as the caudate nucleus, which has the highest concentration of dopamine axon terminals in the brain. Activity there correlated positively with the scores on the subjects' Passionate Love Scales. The results showed that it is better to consider early-stage love a drive, not an emotion. We can have many emotions when we are in love, from euphoria to anxiety, but the essence of love is a drive to be with the other person. It is important that we and others have verified, in additional brain-imaging experiments in New York, London, and Beijing, that the underlying reward system in the brain for early-stage romance is cross-cultural, even though cultural attitudes and ideas about intense romantic love differ widely.[10]

The brain stem reward-and-drive system is also activated when addicts experience a cocaine high.[11] We all know how drugs of abuse produce euphoria and make people do irrational things to get the drug. In this way, romantic love becomes more understandable. It may be a key natural compulsion, a form of positive addiction. Close examinations of the symptoms of compulsive drug use and romantic love show many similarities, especially for the "withdrawal" of heartbreak: both involve euphoria, craving, obsession, and risky behavior to obtain the goal of the other person or the drug.[12] A great hope for the future of this research line is to find ways to help people deal with substance addiction as well as "person addiction" during heartbreak.[13]

The big unanswered question is why we are attracted to, and fall in love with, one particular person and not another. It is clearly not the color of our feathers, although many believe it is the color of nail polish, lip-stick, hair, or sports car! One set of brain-scanning experiments has taken aim at this question.[14] The investigators measured brain activation after a speed-dating session, which enables a rapid evaluation for a potential romantic relationship. The choice the participants made about a person's potential for them based on objective facial attractiveness involved reward regions like the ventral tegmental area. But most important, the idiosyncratic, individual choices ("looks sweet") did not include brain stem reward regions. Rather, they included a part of the neocortex (the prefrontal cortex) that compares us to others and evaluates similarity between us and another person.[15] Of course, no part of the brain does just one thing, but the results suggest that processing our possible *similarities* to another is key to our initial mate-choice decision. This is high-level social cognition, but it is still not necessarily a *conscious* choice. None of the participants in the experiment would have said they were making this comparison.

What about other cognitive processes and individual differences during romance and attachment? Brain-mapping studies show activity changes in the neocortex that reflect our personalities,[16] our empathic tendencies,[17] our satisfaction with a relationship,[18] and how we score on the Storge/Friendship-Based Love Scale in a relationship.[19] There are changes in the neocortex of the brain related to how long we have been in a relationship that are involved in memory, attention, and sense of self.[20] For example, brain activity changes in the inferior frontal gyrus, insular cortex, subcallosal gyrus and prefrontal cortex are related to relationship satisfaction both early and late in relationships, suggesting that empathy, positive experiences, and emotional regulation are important in romance.[21] People who are "Highly Sensitive" or "Negotiators" show more activation than others in the cortical brain regions associated with empathy.[22] Clearly, the neocortex and the cognitive processes it subserves mediate many of our individual differences. But there are no differences in the cortex or brain stem detected so far in romance experiments between men and women and no differences between people with same-sex or opposite-sex partners.[23]

Our brains have neural circuits that help us survive and get our genes into the next generation. Romantic love is part of those circuits. It uses the same cells as those that drive us toward essential, and delicious, food and drink. It uses transmitters and hormones like dopamine, oxytocin, and vasopressin, which are critical to basic functions like reward detection, movement, thirst, and reproduction. Romance and attachment are so basic, so important to our survival, that love is a normal, positive addiction that nature has built into our brains. We need each other, and we spend a lot of time looking for romance, in the same way that we become focused on food when we are hungry. We need each other for protection in this world and for fun, not just for having children and raising a family. The romantic drive is there at all ages, without children, and in same-sex relationships. Intimate romantic relationships enrich, reward, and protect us. We may differ when it comes to the cognitive brain areas that we use in a romantic relationship, but we all share a primitive reward and drive system. We use it to fall in love and pursue another person day in and day out, to form a lasting emotional bond with that person. The brain-scanning data suggest that there is no essential mystery here about our feelings and actions but rather an ancient, evolved natural survival system that is not unique to humans.

NOTES

1. Deborah Anapol, PhD, Psychology Today blog, from her book, *The Seven Natural Laws of Love,* https://www.psychologytoday.com/blog/love-without -limits/201111/what-is-love-and-what-isnt

2. Elizabeth Barrett Browning, "How Do I Love Thee?" https://www.poets.org /poetsorg/poem/how-do-i-love-thee-sonnet-43.

3. D. Tennov, *Love and Limerance: The Experience of Being in Love* (New York: Scarborough House, 1999). Tennov obtained the following descriptions of "limerance" from interviews: infatuation; being in love; involuntary and uncontrollable; intrusive thinking during the day and night about the other person and thus a "fixation"; the other person takes on special meaning; there is craving for the other person, hope for the relationship, and uncertainty; emotional union is craved more than sex, although sex is also craved during "limerance." Tennov devised a questionnaire that she used on four hundred men and women to further verify these feelings.

4. E. Hatfield and S. Sprecher, "Measuring Passionate Love in Intimate Relationships," *Journal of Adolescence* 9 (1986): 383–410. Examples on the Passionate Love Scale, where respondents rate how much they agree with the comment on a scale of 1–6, are the following: 1. Sometimes my body trembles with excitement at the sight of ( ). 2. Sometimes I feel I can't control my thoughts; they are obsessively on ( ). 3. I would rather be with ( ) than anyone else. 4. I yearn to know all about ( ). 5. I will love ( ) forever.

5. C. Hendrick, S. S. Hendrick, and A. Dicke, "The Love Attitudes Scale: Short

Form," *Journal of Social and Personal Relationships* 15 (1998): 147–159. Examples on the Love Attitudes Scale: Eros: My partner and I really understand each other. Ludus: I have sometimes had to keep my partner from finding out about other partners. Storge: Our friendship emerged gradually over time. Pragma: A main consideration in choosing my partner was how he/she would reflect on my family. Mania: I cannot relax if I suspect that my partner is with someone else. Agape: I am usually willing to sacrifice my own wishes to let my partner achieve his/hers.

6. H. E. Fisher: "Lust, Attraction and Attachment in Mammalian Reproduction," *Human Nature* 9 (1998): 23–52, and *Anatomy of Love* (New York: Norton, 2016).

7. Fisher: Lust, Attraction and Attachment in Mammalian Reproduction" and *Anatomy of Love.*

8. A. Aron, H. Fisher, D. J. Mashek, G. Strong, H. Li, and L. L. Brown, "Reward, Motivation, and Emotion Systems Associated with Early-Stage Intense Romantic Love," *Journal of Neurophysiology* 94 (2005): 327–337.

9. Participants in the experiments were instructed to think romantic thoughts when they looked at the face of their beloved. The control for this study was a picture of the face of a familiar person without any emotional attachments, like a fellow student in a class. Thoughts for the familiar neutral person were everyday events. There was also a count-back task (count back by 7's from a large number like 1011) between conditions to prevent spillover of feelings and activations from the time spent looking at the romantic partner from influencing thoughts about the merely familiar person.

10. A. Bartels and S. Zeki, "The Neural Correlates of Maternal and Romantic Love," *Neuroimage* 21 (2004): 1155–1166; X. Xu, A. Aron, L. Brown, G. Cao, T. Feng, and X. Weng, "Reward and Motivation Systems: A Brain Mapping Study of Early-Stage Intense Romantic Love in Chinese Participants," *Human Brain Mapping* 32 (2011): 249–257.

11. H. C. Breiter, R. L. Gollub, R. M. Weisskoff, D. N. Kennedy, N. Makris, J. D. Berke, J. M. Goodman, H. L. Kantor, D. R. Gastfriend, J. P. Riorden, R. T. Mathew, B. R. Rosen, and S. E. Hyman, "Acute Effects of Cocaine on Human Brain Activity and Emotion," *Neuron* 19 (1997): 591–611.

12. H. E. Fisher, L. L. Brown, A. Aron, G. Strong, and D. Mashek, "Reward, Addiction, and Emotional Regulation Systems Associated with Rejection in Love," *Journal of Neurophysiology* 104 (2010): 51–60; H. E. Fisher, X. Xu, A. Aron, and L. L. Brown, "Intense, Passionate, Romantic Love: A Natural Addiction? How the Fields That Investigate Romance and Substance Abuse Can Inform Each Other," *Frontiers in Psychology* 7 (2016): 687.

13. Bartels and Zeki, "The Neural Correlates of Maternal and Romantic Love"; Xu, Aron, Brown, Cao, Feng, and Weng, "Reward and Motivation Systems."

14. J. C. Cooper, S. Dunne, T. Furey, and J. P. O'Doherty, "Dorsomedial Prefrontal Cortex Mediates Rapid Evaluations Predicting the Outcome of Romantic Interactions," *Journal of Neuroscience* 32 (2012): 15647–15656.

15. J. P. Mitchell, C. N. Macrae, and M. R. Banaji, "Dissociable Medial Prefrontal Contributions to Judgments of Similar and Dissimilar Others," *Neuron* 50 (2006): 655–663.

16. L. L. Brown, B. Acevedo, and H. E. Fisher, "Neural Correlates of Four Broad Temperament Dimensions: Testing Predictions for a Novel Construct of Personality," *PLOS One* 8 (2013): e78734.

17. B. P. Acevedo, E. N. Aron, A. Aron, M. D. Sangster, N. Collins, and L. L. Brown, "The Highly Sensitive Brain: An fMRI Study of Sensory Processing Sensitivity and Response to Others' Emotions," *Brain and Behavior* 4 (2014): 580–594.

18. B. Acevedo, A. Aron, H. E. Fisher, and L. L. Brown, "Neural Correlates of Marital Satisfaction and Well-Being: Reward, Empathy and Affect," *Clinical Neuropsychiatry* 9 (2012): 20–31.

19. B. Acevedo, A. Aron, H. Fisher, and L. L. Brown, "Neural Correlates of Long-Term Intense Romantic Love," *Social Cognitive Affective Neuroscience* 7 (2012): 145–159.

20. Deborah Anapol, https://www.psychologytoday.com/blog/love-without-limits/201111/what-is-love-and-what-isnt; X. Xu, L. L. Brown, A. L. Aron, G. Cao, T. Feng, B. Acevedo, and X. Weng, "Regional Brain Activity during Early-Stage Intense Romantic Love Predicted Relationship Outcomes after 40 Months: An fMRI Assessment, *Neuroscience Letters* 526 (2012): 33–38.

21. Acevedo, Aron, Fisher, and Brown, "Neural Correlates of Long-Term Intense Romantic Love"; Xu, Brown, Aron, Cao, Feng, Acevedo, and Weng, "Regional Brain Activity during Early-Stage Intense Romantic Love Predicted Relationship Outcomes after 40 Months"; S. Zeki and J. P. Romaya, "The Brain Reaction to Viewing Faces of Opposite- and Same-Sex Romantic Partners," *PLOS One* 5 (2010): e15802. The investigators found no differences among men, women, gays, or heterosexuals, but future studies with more subjects may find some small differences between men and women in cortical regions. The differences in personal styles are probably greater than gender differences within any group, however.

22. Acevedo, E. N. Aron, A. Aron, Sangster, Collins, and Brown, "The Highly Sensitive Brain"; Acevedo, Aron, Fisher, and Brown, "Neural Correlates of Marital Satisfaction and Well-Being."

23. Zeki and Romaya, "The Brain Reaction to Viewing Faces of Opposite- and Same-Sex Romantic Partners."

Human Sexual Orientation Is Strongly Influenced by Biological Factors

David J. Linden

MANY YEARS AGO, a friend asked me, "Can you recall when you first decided to be heterosexual?" After thinking about it, I responded that it didn't feel like a decision at all. From the very first moments when I began to have sexual or romantic feelings, even before puberty, they were directed toward females. Years later, when I was in college, I had a lot of gay and bisexual friends who said that I should try dating men to see if I might like it. So in the spirit of experimental science, I did. But unlike my experiences with women, I never found men sexually compelling— they just didn't sound or feel or smell right—and I concluded that in my heart of hearts, I was simply straight.

We use the terms "heterosexual," "homosexual," and "bisexual" as categories of what we have come to call "sexual orientation." It's a crude measure—it conflates love with lust and fantasy with real-world behavior. And assigning sexual orientation is an attempt to pin down feelings that may change over time.[1] That said, for most adults these simple categories of sexual orientation—gay/straight/bi—are useful and reasonably stable, and they capture some important aspects of human sexuality. The best estimates from several surveys done in the United States and Europe indicate that approximately 3 percent of men and 1 percent of women identified as homosexual, about 0.5 percent of men and 1 percent of women as bisexual, with the remainder heterosexual.[2]

Most people—gay, straight, or bi—when asked to recall how they came to choose their sexual orientation, will respond as I did: "It didn't feel like a choice but rather like a deep compulsion that was evident as soon as I became aware of my sexual feelings." Indeed, in one survey only about 4 percent of all gay and bi men reported that they chose their sexual orientation.[3] It's interesting that this number was somewhat higher among lesbians and bi women, at 15 percent.[4] This result is consistent with other lines of evidence suggesting that sexual orientation in women is somewhat more malleable than it is in men.[5]

Might sexual orientation be determined by early social experience, before overt sexual feelings emerge? Sigmund Freud, the founder of psychoanalysis, held that male homosexuality originated in childhood and resulted from close-binding mothers and hostile fathers (he had much less to say about female homosexuality). Others have claimed that both male and female homosexuality can result from childhood physical or sexual abuse. In truth, there's no clear evidence to support a causal link between any of these early social experiences and adult sexual orientation.[6] Moreover, in many cases there isn't even clear evidence to support a correlation. For example, surveys comparing straight women and lesbians have differed, with some reporting a higher incidence of prior childhood physical and sexual abuse among lesbians and others failing to find a difference.[7] Furthermore, children who are raised by single mothers or lesbian couples are no more or less likely to be straight than those raised by straight couples.[8] It is remarkable that there is no clear evidence to establish that *any* practices of child-rearing influence adult sexual orientation.[9]

Is there a genetic component to sexual orientation? If you are a woman with a lesbian sister, then your chance of being a lesbian is about 16 percent (compared to 1 percent in the general population), and if you're a man with a gay brother, then your chance of being gay is about 22 percent (versus 3 percent in the general population).[10] These statistics tell us that sexual orientation clusters in families, but they do not tell us why. For example, two brothers share, on average, 50 percent of their genes with each other, but they also tend to share a similar upbringing. So if Freud was right and a close-binding mother causes a boy to become gay, then this effect could be found in two (or more) brothers raised together.

One way to disentangle genetics from upbringing is to analyze same-sex twins. Fraternal twins (also called dizygotic twins) are no more genetically related to each other than same-sex singleton siblings; they also share about 50 percent of their genes. Monozygotic twins (sometimes called identical twins) share 100 percent of their genes, as well as a similar upbringing (except for the rare cases of twins raised apart; see the essay by Jeremy Nathans for further discussion). So if sexual orientation had no genetic component, we'd expect that the percentage of twin pairs where both are gay would be roughly the same for dizygotic and monozygotic twins. Conversely, if sexual orientation were entirely determined by genes, then every gay monozygotic twin would have a gay twin sibling. The best estimates to date, from a population of 3,826 randomly selected twin pairs in Sweden, indicate that sexual orientation in women is about 20 percent determined by genes, and in men, it's about 40 percent determined by genes.[11] These percentages should be taken as approximate figures. Nonetheless, there is a clear general conclusion: genetics is an important factor in determining sexual orientation, but it is far from the whole story and it is a somewhat stronger effect in men than in women.[12] It's important to remember that these estimates of heritability are measures for populations, not individuals. It may be that there are some individual women and men who carry gene variants such that their sexual orientation is entirely genetically determined and others for whom sexual orientation has no genetic contribution whatsoever.

There is no single gene that determines human sexual orientation. Rather, many different genes each appear to make a small contribution, and at present, we do not have a list of these genes. This is not surprising. Many complex behavioral traits, like shyness or general intelligence or risk taking, have significant genetic components yet are not determined by a single gene. Even a straightforward physical trait like height, which is about 80 percent genetically determined, is influenced by over two hundred different genes involved in the growth of cartilage, bone, muscle, etc.[13] The lack of a single gene for sexual orientation in humans is not an argument against a genetic contribution to this trait.

If child rearing has little or no effect on sexual orientation and genes have only a partial effect, then why are some people straight while others are bi or gay? At present, the most promising insights come from biological effects that are independent of genes, most notably exposure to sex hormones (and possibly signaling molecules of the immune system) in utero and in early postnatal life. For cisgendered people, males typically have an X and a Y chromosome, whereas females usually have two X chromosomes. One crucial gene on the Y chromosome, called SRY, codes for an important protein that, by affecting the activation of other genes, instructs male development beginning early in embryonic development. In the presence of the SRY gene product, two small blobs of embryonic tissue are instructed to become testes, which secrete the sex hormone testosterone. Testosterone then has widespread effects throughout the body and is the key signal driving male-typical development of everything from the genitalia to the Adam's apple to the brain. In females, other genes drive these same blobs of embryonic tissue to become ovaries, which secrete the key hormones estrogen and progesterone. It is important that while testosterone is present in two different surges starting in early fetal life, the secretion of estrogen is suppressed from the period shortly after birth to puberty. This means that during certain critical stages of development the main sex hormone difference is higher levels of circulating testosterone (and some testosterone-like hormones, collectively called androgens) in most males and lower levels in most females.[14]

Here's one useful hypothesis: when female fetuses and babies are exposed to higher levels of testosterone, their brains are partially masculinized, and this increases the chance that, later in life, they will become sexually attracted to women. Similarly, when male fetuses and babies are exposed to lower levels of testosterone, their brains become partially feminized, thereby increasing the chance that they will eventually become sexually attracted to men. There are several lines of evidence in support of this idea.[15] Females with a condition called congenital adrenal hyperplasia have increased testosterone levels in fetal life. Even when they are treated with testosterone-blocking drugs starting at birth, their brains appear to have been partially masculinized. About 21 percent of women with congenital adrenal hyperplasia report sexual attraction to women (compared to 1.5 percent in the general female population).[16] This finding is consistent with some experiments in laboratory animals: when guinea pigs, rats, or sheep receive treatments that boost fetal testosterone signaling, the females grow up to display male-typical sexual behaviors. That is, they mount females and fail to exhibit a posture called lordosis, which encourages males to mount them. Similarly, treatments that attenuate testosterone signaling in developing males reduced male-typical sexual behavior in adult rats and sheep.

What differences can we observe in the brains of men and women? Can we use those differences to test the hypothesis that the brains of gay men are more likely to be partly feminized and the brains of lesbians are more likely to be partly masculinized? In humans, we're mostly limited to analysis of autopsy tissue. These investigations have revealed that there are areas of the brain that appear to vary in size between straight men and straight women. These include a portion of the hypothalamus called INAH3, which is larger in straight men, and a bundle of fibers that connects one side of the brain to the other, called the anterior commissure, which is larger in straight women. While there have been some well-publicized reports suggesting that the size of the anterior commissure and INAH3 are more femalelike in gay men,[17] there have yet to be clear, independent replications of these findings.[18]

It is important to realize that measuring the size of a particular brain region in human autopsy tissue is an extremely crude measure. The functions of brain regions, including those that influence sexual orientation, are determined by many subtle factors such as the chemical, electrical, and micro-anatomical properties of neurons and their responses to experience. Profound differences in neural function can be produced by changes that would never be detected by measuring the sizes of brain regions. It may be, for example, that one key difference between the brains of straight women and lesbians is the function of a particular protein, like a voltage-sensitive potassium channel, that influences the electrical behavior of certain neurons in a brain circuit that influences sexual and gender-typical behavior.

From the earliest portion of childhood there are, on average, behavioral differences between girls and boys. For example, boys are more likely to engage in rough-and-tumble play and to interact with inanimate object toys, while, on average, girls are more likely to play less aggressively and to choose dolls and animals for their toys.[19] When populations of girls and boys were evaluated for gender-typical behavior and then followed to adulthood, an amazing result emerged. Boys who showed highly female-typical behaviors early on were much more likely to become sexually attracted to men as adults (75 percent, compared with 3.5 percent of the general population), and girls who showed male-typical behaviors were far more likely to become sexually attracted to women as adults (24 percent, compared with 1.5 percent of the general population).[20] However, this isn't a universal result: not all tomboys develop sexual attraction to women, and not all effeminate boys become sexually attracted to men. And of course, in adults, not all lesbians are mannish, and not all gay men are effeminate. Yet these findings are striking and suggest a general explanation: sexual orientation is just one aspect of variation in brain function that produces a set of behaviors that can be more or less gender-typical. For example, gender nonconforming girls are more likely to engage in rough-and-tumble play, are less likely to engage in cooperative social play, and are more likely to become sexually attracted to women as they grow up. The most likely explanation is that some combination of genes, fetal signals such as circulating sex hormones and perhaps other biological factors we do not yet understand, influence the relevant brain circuits for gender-typical behavior, of which sexual orientation is just one part of the package.

It is crucial that this model also provides a possible explanation for why Freud and others got it so wrong about child rearing. On average, gay men *do* tend to report stronger childhood bonds with their mothers and weaker bonds with their fathers than straight men.[21] Freud, in considering the childhood reports of his gay male clients had the correlation right but had the causality wrong. It's not that child rearing strongly influences sexual orientation. Rather, variation in brain circuits that influence gender-typical behaviors, first evident in young children, affect how parents and other adults respond to them.

NOTES

1. Some people are offended by this lack of subtlety. They adopt newer terms like "pansexual" (not limited in sexual choice with regard to biological sex, gender, or gender identity); "demisexual" (sexual attraction only to people with whom one has an emotional bond); or "heteroflexible" (minimal homosexual activity in an otherwise primarily heterosexual orientation), or they even refuse any single-word descriptor of their sexual or romantic feelings. An interesting attempt to create a more subtle and inclusive set of descriptors of sexual feelings that accounts for variations in gender identity may be found in S. M. van Anders, "Beyond Sexual Orientation: Integrating Gender/Sex and Diverse Sexualities via Sexual Configurations Theory,"

Archives of Sexual Behavior 44 (2015): 1177–1213. In this essay, I haven't explored the crucial issues of gender identity and transexuality, not because they are unimportant but because they deserve much more consideration than I can provide in this short-essay format. Clearly, how you identify your own sex and whether it corresponds to your assigned gender is foundational for the subsequent issue of sexual orientation.

2. There have now been several large-scale anonymous surveys of sexual orientation performed using random sampling. While the results from these surveys have been fairly consistent, the caveat remains that people do not always answer surveys honestly, so there may be some systematic error in the results. In addition, these surveys have mostly been conducted in higher-income English-speaking countries, and the results might be somewhat different in other populations. The following survey, conducted in the United States, is representative of this group: E. O. Laumann, J. H. Gagnon, R. T. Michael, and S. Michaels, *The Social Organization of Sexuality: Sexual Practices in the United States* (Chicago: University of Chicago Press, 1994). It is interesting that more recent surveys, conducted in times when societal approval of homosexuality and bisexuality has generally increased, have not shown a higher incidence of homosexuality and bisexuality in their responses.

3. J. Lever, "Sexual Revelations: The 1994 Advocate Survey of Sexuality and Relationships: The Men," *The Advocate*, August 23, 1994, pp. 17–24.

4. J. Lever, "Lesbian Sex Survey," *The Advocate*, August 22, 1995, pp. 21–30. Simon LeVay has stated the matter succinctly: "If their sexual orientation was indeed a choice, gay people should remember having made it. But, by and large, they don't." S. LeVay, *Gay, Straight, and the Reason Why: The Science of Sexual Orientation* (Oxford: Oxford University Press, 2010), p. 41. For a nice explanation by some who state that they have chosen their homosexual orientation, see http://www.queerbychoice.com/.

5. L. M. Diamond, *Sexual Fluidity: Understanding Women's Love and Desire* (Cambridge, MA: Harvard University Press, 2008).

6. Even when positive associations are found between childhood maltreatment and same-sex attraction in adulthood, the devil is in the details when ascribing causality. Roberts et al. nicely explain how there are different potential explanations that are not easily untangled: (1) childhood abuse increases the likelihood of same-sex attraction developing and being expressed in adulthood; (2) nascent homosexuality causes childhood maltreatment; (3) reported differences in childhood maltreatment result not from a higher incidence of abuse but rather from a higher incidence of recall among gay men and lesbians. A. L. Roberts, M. M. Glymour, and K. C. Koenen, "Does Maltreatment in Childhood Affect Sexual Orientation in Adulthood?" *Archives of Sexual Behavior* 42 (2013): 161–171.

7. J. C. Brannock and B. E. Chapman, "Negative Sexual Experiences with Men among Heterosexual Women and Lesbians," *Journal of Homosexuality* 19 (1990): 105–110; J. P. Stoddard, S. L. Dibble, and N. Fineman, "Sexual and

Physical Abuse: A Comparison between Lesbians and Their Heterosexual Sisters," *Journal of Homosexuality* 56 (2009): 407–420.

8. F. L. Tasker and S. Golombok, *Growing Up in a Lesbian Family: Effects on Child Development* (New York: Guilford, 1997); R. Green, J. B. Mandel, M. E. Hotvedt, J. Gray, and L. Smith, "Lesbian Mothers and Their Children: A Comparison with Solo Parent Heterosexual Mothers and Their Children," *Archives of Sexual Behavior* 7 (1986): 175–181; C. J. Patterson, *Lesbian and Gay Parents and Their Children: Summary of Research Findings* (Washington, D.C.: American Psychological Association, 2005).

9. Perhaps the most dramatic cases to illustrate the independence of early social experience and sexual orientation are those in which boys born with penile malformation have received gender reassignment surgery as infants and have been raised as girls from the time of birth. The theory that drove these ill-fated decisions held that infants are a blank slate, so chromosomal males could be raised to feel female and be attracted to men as adults. In fact, this completely failed. As they grew up, almost all reassigned boys reported that they felt male, and almost all grew to become sexually attracted to women. M. Diamond and H. K. Sigmundson, "Sex Reassignment at Birth: Long-Term Review and Clinical Implications," *Archives of Pediatric and Adolescent Medicine* 151 (1997): 298–304.

10. J. M. Bailey and R. C. Pillard, "A Genetic Study of Male Sexual Orientation," *Archives of General Psychiatry* 48 (1991): 1089–1096; J. M. Bailey and D. S. Benishay, "Familial Aggregation of Female Sexual Orientation," *American Journal of Psychiatry* 150 (1993): 272–277. What remains unclear is whether family clusters occur for opposite-sex siblings: if a woman has a gay brother, does that increase her chance of being sexually attracted to women, and if a man has a lesbian sister, does that increase his chance of being sexually attracted to men? So far, the answers appear to be no.

11. N. Långström, Q. Rahman, E. Carlström, and P. Lichtenstein, "Genetic and Environmental Effects on Same-Sex Sexual Behavior: A Population Study of Twins in Sweden," *Archives of Sexual Behavior* 39 (2010): 75–80. This Swedish study is notable because of its large sample size and random sampling of twins. Another study that used a large randomized population of British female twins yielded a similar estimate for the degree of heritability of sexual orientation in women: 25 percent; A. Burri, L. Cherkas, T. Spector, and Q. Rahman, "Genetic and Environmental Influences on Female Sexual Orientation, Childhood Gender Typicality and Adult Gender Identity," *PLOS One* 6 (2011): e21982. Some previous twins studies had produced higher estimates of heritability for both male and female sexual orientation (around 50 percent for both), but those studies used self-selected volunteers (through advertisements in the lesbian and gay press and at Gay Pride events), and it is possible that such a practice may have biased the results.

12. The pattern of inheritance for sexual orientation remains confusing. Several

family pedigree studies among men have indicated that maternal transmission may be involved, but there was a subsequent large and careful study that failed to replicate this effect. Several investigations have described increased rates of same-sex attraction among nonsibling relatives of lesbians and gay men. These include the daughters, nieces, and female cousins of lesbians and the uncles and male cousins of gay men; A. M. Pattatucci and D. H. Hamer, "Development and Familiality of Sexual Orientation in Females," *Behavioral Genetics* 25 (1995): 407–420; D. H. Hamer, S. Hu, V. L. Magnuson, N. Hu, and A. M. Pattatucci, "A Linkage between DNA Markers on the X Chromosome and Male Sexual Orientation," *Science* 261 (1993): 321–327.

13. A. R. Wood et al., "Defining the Role of Common Variation in the Genomic and Biological Architecture of Adult Human Height," *Nature Genetics* 46 (2014): 1173–1186.

14. Females at all stages of development and in adulthood are not completely bereft of androgen signaling. The adrenal glands secrete low levels of testosterone and testosterone-like molecules such as dihydrotestosterone and androstenedione.

15. This hypothesis may turn out to be both true and incomplete. It is likely that other hormones in addition to testosterone are important in influencing sexual orientation.

16. H. F. Meyer-Bahlburg, C. Dolezal, S. W. Baker, and M. I. New, "Sexual Orientation in Women with Classical or Non-Classical Congenital Adrenal Hyperplasia as a Function of Degree of Prenatal Androgen Excess," *Archives of Sexual Behavior* 37 (2008): 85–99.

17. L. S. Allen and R. A. Gorski, "Sexual Orientation and the Size of the Anterior Commissure in the Human Brain," *Proceedings of the National Academy of Sciences of the USA* 89 (1992): 7199–7202; S. LeVay, "A Difference in Hypothalamic Structure between Heterosexual and Homosexual Men," *Science* 253 (1991): 1034–1037.

18. W. Byne, S. Tobet, L. A. Mattiace, M. S. Lasco, E. Kemether, M. A. Edgar, S. Morgello, M. S. Buchsbaum, and L. B. Jones, "The Interstitial Nuclei of the Human Anterior Hypothalamus: An Investigation of Variation with Sex, Sexual Orientation, and HIV Status," *Hormones and Behavior* 40 (2001): 86–92; M. S. Lasco, T. J. Jordan, M. A. Edgar, C. K. Petito, and W. Byne, "A Lack of Dimorphism of Sex or Sexual Orientation in the Human Anterior Commissure," *Brain Research* 936 (2002): 95–98.

19. Many lines of evidence suggest that biological factors play an important role in influencing children's gender development. For example, on average, newborn girls prefer to look at faces, whereas newborn boys prefer to look at mobiles; J. Connellan, S. Baron-Cohen, S. Wheelwright, A. Batki, and J. Ahluwalia, "Sex Differences in Human Neonatal Social Perception," *Infant Behavior and Development* 23 (2001): 113–118.

20. R. Green, *The "Sissy-Boy Syndrome" and the Development of Homosexuality*

(New Haven: Yale University Press, 1987); K. D. Drummond, S. J. Bradley, M. Peterson-Badali, and K. J. Zucker, "A Follow-Up Study of Girls with Gender Identity Disorder," *Developmental Psychology* 44 (2008): 34–45.

21. R. A. Isay, "Gender Development in Homosexual Boys: Some Developmental and Clinical Considerations," *Psychiatry* 62 (1999): 187–194.

DECIDING

Deep Down, You Are a Scientist

Yael Niv

YOU MAY NOT KNOW IT, but deep down you are a scientist. To be precise, your brain is a scientist—and a good one, too: the kind of scientist that makes clear hypotheses, gathers data from several sources, and then reaches a well-founded conclusion.

Although we are not aware of the scientific experimentation occurring in our brain on a momentary basis, the scientific process is fundamental to how our brain works. This process involves three key components. First: hypotheses. Our brain makes hypotheses, or predictions, all the time. Every movement that we make involves predicting: Where will my arm end up if I engage this muscle? How heavy is the cup of coffee that I am planning to grasp and bring to my mouth? Etc.[1] A stark example is the familiar experience of going up the stairs in the dark (or while reading email on your phone) and almost falling because you expected, predicted, that there would be one more stair. You could have simply lifted your leg and decided whether there was a stair or not based on sensory feedback (Do I feel a hard surface or not?), but anyone who has traveled up a flight of stairs in pitch dark knows that this is extremely slow and attention-demanding. Our normal stair climbing is quick and effortless thanks to accurate predictions.

This is also how we learn from experience: we don't just wait for things to happen in order to learn, but rather we make a prediction about

227

what will happen and learn only if our prediction is wrong.[2] Every time you cross the street, you predict whether an approaching car will make it to the crosswalk before you make it to the other side of the street. This is not a simple prediction; it relies on inferring the distance and speed of the car from visual input, as well as having a good idea of how fast you walk and how wide the street is. Nevertheless, young kids can master this complex inference and prediction. If you have taught young kids to cross the street, you may have noticed, however, that they are much more cautious than (most) adults. This is because they have yet to refine their mental model of the speed of cars and how long it takes to cross a street. They will do this through trial and error by observing differences between their predictions and reality; every time they cross the street, although they will have crossed successfully, as their brain predicted, their brain will automatically register small discrepancies between prediction and reality: "That car is still very far from me, even though I am close to the other sidewalk." They will learn to adjust their predictions accordingly and, over time, will cross the street more like an adult.

We know, from over a century of research about learning from experience with the world, that animals, from snails and bees[3] to monkeys and humans,[4] all learn by making predictions and then comparing these predictions to reality as it unfolds. This is called "error-driven learning": you don't learn just from what happens; you specifically learn from your mistakes in predicting what will happen.

The second component of good scientific work is gathering data—testing your hypothesis by comparing it to evidence. As neuroscientists, we can gather data to test our theories about how the brain works from several sources—for example, behavior, invasive recordings of the activity of single cells in the brain, and noninvasive imaging of overall activity in large areas of the brain. Because each type of measurement gives only partial information about how the brain works, to make solid conclusions we are taught to combine information from as many sources as possible. You can call this "corroborating evidence," and it is not only the mainstay of scientific discovery, but is also important in fields ranging from journalism to art history (you would not make conclusions about an artist's style from only one painting, or even only from paintings of that artist, without comparison to other contemporaries).

Our brain seeks corroborating evidence automatically: it optimally combines information from several sources in order to understand the world.[5] These sources are our senses. Have you ever felt that you can better hear what someone is saying when you have a line of sight to the person's face? That is because your brain is combining vision (yes, you can lip-read!) and sound to interpret speech, and this is most important when there are distractors around (like other people talking or some background noise).[6] Another example is hammering a nail into a wall. We intuitively know that the best way to avoid hitting our thumb (or the wall) is by having it be our thumb (not someone else's!) holding the nail. Why is that? If someone else held the nail, we would surely avoid pain to ourselves. Yet we are not as confident aiming the hammer if we are not holding the nail because vision (seeing where the nail is) is not enough. We also use the sense of proprioception—our internal knowledge, based on sensors in our joints, of where our limbs are in three-dimensional space. By holding the nail ourselves, we can combine proprioception and vision to accurately aim.

Finally, after making precise, well-founded predictions and gathering data from all available sources, a scientist must interpret the empirical observations. So does the brain. The world is inherently ambiguous, allowing multiple interpretations of our perceptual input at any point in time. Imagine passing by your kitchen at night and seeing light coming from the window. Is this the light inside the room, reflecting back from the glass pane, or a burglar outside shining a flashlight into your house? To interpret the sensory information, your brain will optimally combine your prior beliefs about each of these events (burglars are rare; the windows often reflect light) with the likelihood of the sensory information under each interpretation (At what angle is the light reflected? Is this the same angle you have witnessed many times? That is, how likely is it that you would perceive this exact scene if it were a reflection versus if it were a burglar?) to arrive at a split-second decision.[7]

It is important to realize that our perceived reality is subjective—it is interpreted—rather than an objective image of the world out there. And in some cases this interpretation can break down. For instance, in schizophrenia, meaningless events and distractors can take on outsized meaning in subjective interpretation, leading to hallucinations, delusions, and paranoia (that faint sound is not just steam in the radiator but rather someone trying to convey a message to me, or aliens spying on one's actions).

Our memories are similarly a reflection of our own interpretations rather than a true record of events. This holds implications for the reliability of memory in witness testimony or in an argument with one's partner.[8]

In essence, our brain is always striving to understand the "truth"— exactly what is out there. But our perception, far from a simple recording of objective reality, is rather an attempt to divine the causal structures that underlie our sensory inputs: what is the simplest "theory" that would explain what we hear, see, or smell? As many perceptual illusions attest, we don't really see the world as it is, but rather, what we perceive is an interpretation, the best story we can tell that would make sense of all the data so far. Just like a scientist.

NOTES

1. M. Kawato, "Internal Models for Motor Control and Trajectory Planning," *Current Opinion in Neurobiology* 9 (1999): 718–727.
2. R. A. Rescorla and A. R. Wagner, "A Theory of Pavlovian Conditioning: Variations in the Effectiveness of Reinforcement and Nonreinforcement," in *Classical Conditioning II: Current Research and Theory,* ed. A. H. Black and W. F. Prokasy (New York: Appleton-Century-Crofts, 1972), pp. 64–99; Y. Niv and G. Schoenbaum, "Dialogues on Prediction Errors," *Trends in Cognitive Science* 12 (2008): 265–272.
3. P. R. Montague, P. Dayan, C. Person, and T. J. Sejnowski, "Bee Foraging in Uncertain Environments Using Predictive Hebbian Learning," *Nature* 377 (1995): 725–728.
4. M. J. Frank, L. C. Seeberger, and R. C. O'Reilly, "By Carrot or by Stick: Cognitive Reinforcement Learning in Parkinsonism," *Science* 306 (2004): 1940–1943.
5. K. P. Körding and D. M. Wolpert, "Bayesian Integration in Sensorimotor Learning," *Nature* 427 (2004): 244–247.
6. H. McGurk and J. MacDonald, "Hearing Lips and Seeing Voices," *Nature* 264 (1976): 746–748.
7. D. C. Knill and W. Richards, *Perception as Bayesian Inference* (Cambridge: Cambridge University Press, 1996); B. A. Olshausen, "Perception as an Inference Problem," in *The Cognitive Neurosciences V,* ed. V. M. Gazzaniga and R. Mangun (Cambridge, MA: MIT Press, 2013).
8. E. F. Loftus, "Eavesdropping on Memory," *Annual Review of Psychology* 68 (2017): 1–18.

Studying Monkey Brains Can Teach Us about Advertising

Michael Platt

ASK ANY AD EXECUTIVE, and he or she will tell you that sex sells. The use of sex to market products is not new. In 1871, the Pearl Tobacco brand printed a poster with a nude woman accompanying the brand name, and a few years later W. Duke & Sons included trading cards with pictures of sexually appealing women in its cigarette packs. In neither case were sexy images related to the nature or quality of the product, but nevertheless sales increased. Fast-forward a century and a half, and it seems impossible to escape from sexual imagery in advertising. But why does sex sell? Do sexy images tap into our base urges and drives, somehow motivating us to buy products?[1] Or do marketers merely exploit—and perpetuate—our cultural obsession with sexual imagery, which is learned and propagated through media?[2]

A related practice is the use of high-profile celebrities to market products. Today, entertainers, including actors, musicians, and athletes, are the star attractions for advertising campaigns, but even powerful people like politicians and business leaders can help sell products. But why does celebrity sell? Given that celebrities often don't even use the products they're advertising, their endorsements are irrelevant for consumer experience. One explanation cited by market researchers is that people want to be more like celebrities—in terms of beauty, wealth, and power—and think that using the same products will allow them to do so.[3]

231

My colleagues and I recently offered a new explanation: sex and celebrity sell because our brains are fine-tuned to prioritize social information.[4] Social life is a major challenge that has resulted in the evolution of brain circuits specialized for identifying others, remembering them, and making inferences about their mental states in order to predict their behavior and make good social decisions.[5] We're not alone in how strongly our brains are tuned to our social worlds. Many species of monkeys and apes—our closest living relatives—also live in large, complex, dynamic societies.[6] Behavioral studies show these primates also identify others, keep track of prior encounters, empathize with friends and relatives, and make inferences about others' mental states.[7]

Paying close attention to social cues can help you make better decisions. From an evolutionary perspective, these cues include whether a member of the opposite sex might be a good potential mate and the presence of high-status individuals who might drive you away from a desired mate or nutritious food. In fact, both men and male monkeys are exquisitely sensitive to cues that indicate female fertility. Men rate women who are ovulating as more attractive than those who are not[8] and tip more for lap dances by women who are ovulating.[9] Similarly, male rhesus macaques prefer to view images of females that have been artificially reddened,[10] because face and hindquarter reddening predicts ovulation and sexual receptivity.[11]

Although based on fewer studies, available data indicate that women and female monkeys are also sensitive to cues about male quality. Women's preferences shift toward more masculine faces—with broader jaws and wider-set but smaller eyes[12]—during ovulation,[13] and female macaques tend to mate with higher-ranking males when they are ovulating[14] and also prefer males with reddened faces caused by a surge in testosterone in the breeding season.[15] Other studies found that both people and monkeys pay more attention to individuals who are high status and are more likely to follow their gaze.[16]

These findings endorse the idea that cues to mate quality and status are useful for guiding behavior and are prioritized by our brains. We tested this idea experimentally in people and monkeys. We developed a behavioral test, which we dubbed the "pay-per-view test"; it measured how much individuals value visual images without their being aware of it.[17] On each trial, participants chose between two options on a computer screen. Choosing either resulted in a predictable food or juice reward for monkeys and a predictable cash reward for people. One of the options—the "social" option—also displayed a picture that was irrelevant to the food or cash payout, whereas the other option—the "blank" option—did not. On some trials, the picture option paid more, and on some trials less, than the blank option. Across many trials we randomly varied which pictures were shown. For monkeys, pictures included faces of high-status males, low-status males, females, and male and female genitalia. For people, pictures were of members of the opposite sex who varied in physical attractiveness according to ratings by a separate group of participants.

By computing how often participants chose each picture as a function of its payout, we could infer how much they would pay to see each type of picture. This "willingness to pay" is a classic measure of economic value.[18] We found male college students paid about 0.25 cents more to view an attractive woman than an average woman. Female students, by contrast, were less motivated to see hot men. Monkeys also valued sex and status.[19] Both males and females paid (by accepting less food or juice) to see images of monkey genitalia and faces of high-status males but had to be paid to see faces of low-status males. We validated our willingness-to-pay measure in humans by showing that it predicted how long participants would wait for an image and how hard they would work to keep it displayed.

These findings indicate that information about sex and status is valuable and can substitute for other rewards like food, juice, and money. The parallels between monkeys and humans are so striking that they suggest that attentional prioritization of information about sex and status is accomplished by brain mechanisms shared by our primate cousins and us. To test this idea, we used functional magnetic resonance imaging to scan the brains of avowed heterosexual male students while they viewed female faces of varying attractiveness and while they viewed money either deposited or withdrawn from their study payment.[20] These scans revealed that a network of brain areas previously implicated in processing rewards—including the orbitofrontal cortex, ventromedial prefrontal cortex, and medial and ventral striatum—was strongly activated by the sight of attractive female faces and that neural activity increased with increasing attractiveness. We found the same effects for monetary rewards and losses, suggesting this network signals the abstract value of both social and monetary rewards. These signals represent a biological correlate of the economic concept of "utility"—one's internal desire for or satisfaction with a good or service—thought to underlie decisions.[21]

To determine the physiological basis of these signals, we used miniature electrical sensors to measure the activity of individual brain cells in these same areas in monkeys. Some cells responded strongly with electrical activity when male monkeys chose to see female genitalia, a high-status male face, or a large juice reward, but responded less when monkeys chose low-status faces or small juice rewards.[22] We also identified neurons that responded to images of faces and genitals but not juice. These findings suggest there are neurons specialized for identifying and prioritizing important social information within the brain's reward system.[23]

Neurons in the parietal cortex, an area important for paying attention, signaled the total value of an option based on both its social importance and its juice payout.[24] Convergence of social and nonsocial value signals in these neurons may allow us to adroitly navigate complex environments with many different stimuli competing for our attention.

Can these discoveries help explain the power of sex and status in advertising? We hypothesize that the brain mechanisms that prioritize social information are also activated by advertisements that associate sex or status with specific brands or products, and this activation may bias preferences toward the product.[25] To test this idea, we ran a social advertising campaign with rhesus macaques by exposing monkeys to logos of household brands (for example, Nike, Adidas, Domino's, Pizza Hut) paired with either a social image (that is, female genitalia, high-status male face, low-status male face) or the same image with the pixels rearranged to make it unrecognizable while retaining the same brightness, contrast, and color, which are other salient cues that can draw attention to a stimulus. Monkeys were given a sweet treat for touching the screen following the ad. They were then offered choices between brands that either had been paired with a social image or its scrambled version.

Our advertising campaign was remarkably effective. Monkeys developed preferences for brands associated with sex and status. Both males and females preferred brands paired with sexual cues and the faces of high-status monkeys. These findings endorse the hypothesis that the brain mechanisms that prioritize information about sex and status shape consumer behavior today, to the advantage of marketers and, perhaps, our own dissatisfaction.

NOTES

1. V. Griskevicius and D. T. Kenrick, "Fundamental Motives: How Evolutionary Needs Influence Consumer Behavior," *Journal of Consumer Psychology* 23 (2013): 372–386.

2. G. Saad, S. Gad, and G. Tripat, "Applications of Evolutionary Psychology in Marketing," *Psychology and Marketing* 17 (2000): 1005–1034. Do sexual images help sell things to women? Increasingly, sexy images of men are used in advertising, but it's not clear who is motivated by those images (i.e., gay men versus straight women). In our work, women are much less moved by sexy images of men; B. Y. Hayden, P. C. Parikh, R. O. Deaner, and M. L. Platt, "Economic Principles Motivating Social Attention in Humans," *Proceedings of the Royal Society, Series B* 274 (2007): 1751–1756.

3. M. Muda, M. Mazzini, M. Rosidah, and P. Lennora, "Breaking through the Clutter in Media Environment: How Do Celebrities Help?" *Procedia—Social and Behavioral Sciences* 42 (2012): 374–382.

4. J. T. Klein, R. O. Deaner, and M. L. Platt, "Neural Correlates of Social Target Value in Macaque Parietal Cortex," *Current Biology* 18 (2008): 419–424; M. Y. Acikalin, K. K. Watson, G. J. Fitzsimons, and M. L. Platt, "What Can Monkeys Teach Us about the Power of Sex and Status in Advertising?" *Marketing Science,* under review.

5. S. W. C. Chang, L. J. N. Brent, G. K. Adams, J. T. Klein, J. M. Pearson, K. K. Watson, et al., "Neuroethology of Primate Social Behavior," *Proceedings of the National Academy of Sciences of the USA* 110 (2013): 10387–10394.

6. M. L. Platt, R. M. Seyfarth, and D. L. Cheney, "Adaptations for Social Cognition in the Primate Brain," *Philosophical Transactions of the Royal Society of London, Series B* 371 (2016): 20150096.

7. L. J. N. Brent, S. W. C. Chang, J.-F. Gariépy, and M. L. Platt, "The Neuroethology of Friendship," *Annals of the New York Academy of Sciences* 1316 (2014): 1–17.

8. M. G. Haselton and K. Gildersleeve, "Can Men Detect Ovulation?" *Current Directions in Psychological Science* 20 (2011): 87–92.

9. G. Miller, J. M. Tybur, and B. D. Jordan, "Ovulatory Cycle Effects on Tip Earnings by Lap Dancers: Economic Evidence for Human Estrus?" *Evolution and Human Behavior* 28 (2007): 375–381.

10. C. Waitt, M. S. Gerald, A. C. Little, and E. Kraiselburd, "Selective Attention toward Female Secondary Sexual Color in Male Rhesus Macaques," *American Journal of Primatology* 68 (2006): 738–744.

11. C. R. Carpenter, "Sexual Behavior of Free Ranging Rhesus Monkeys *(Macaca*

mulatta). I. Specimens, Procedures and Behavioral Characteristics of Estrus," *Journal of Comparative Psychology* 33 (1942): 113–142; J. P. Higham, K. D. Hughes, L. J. N. Brent, C. Dubuc, A. Engelhardt, M. Heistermann, et al., "Familiarity Affects the Assessment of Female Facial Signals of Fertility by Free-Ranging Male Rhesus Macaques," *Proceedings of the Royal Society, Series B* 278 (2011): 3452–3458.

12. P. Mitteroecker, S. Windhager, G. B. Muller, and K. Schaefer, "The Morphometrics of 'Masculinity' in Human Faces," *PLOS One* 10 (2015): 2.

13. J. R. Roney and Z. L. Simmons, "Women's Estradiol Predicts Preference for Facial Cues of Men's Testosterone," *Hormones and Behavior* 53 (2008): 14–19.

14. J. H. Manson, "Measuring Female Mate Choice in Cayo Santiago Rhesus Macaques," *Animal Behavior* 44 (1992): 405–416.

15. C. Waitt, A. C. Little, S. Wolfensohn, P. Honess, A. P. Brown, H. M. Buchanan-Smith, et al., "Evidence from Rhesus Macaques Suggests That Male Coloration Plays a Role in Female Primate Mate Choice," *Proceedings of the Royal Society, Series B* 270 (2003): S144–S146.

16. B. C. Jones, L. M. DeBruine, J. C. Main, A. C. Little, L. L. M. Welling, D. R. Feinberg, et al., "Facial Cues of Dominance Modulate the Short-Term Gaze-Cuing Effect in Human Observers," *Proceedings of the Royal Society, Series B* 277 (2010): 617–624; S. V. Shepherd, R. O. Deaner, and M. L. Platt, "Social Status Gates Social Attention in Monkeys," *Current Biology* 16 (2006): R119–R120; M. Dalmaso, G. Pavan, L. Castelli, and G. Galfano, "Social Status Gates Social Attention in Humans," *Biology Letters* 8 (2012): 450–452; M. Dalmaso, G. Galfano, C. Coricelli, and L. Castelli, "Temporal Dynamics Underlying the Modulation of Social Status on Social Attention," *PLoS One* 9 (2014): e93139.

17. R. O. Deaner, A. V. Khera, and M. L. Platt, "Monkeys Pay per View: Adaptive Valuation of Social Images by Rhesus Macaques," *Current Biology* 15 (2005): 543–548; B. Y. Hayden and M. L. Platt, "Gambling for Gatorade: Risk-Sensitive Decision Making for Fluid Rewards in Humans," *Animal Cognition* 12 (2009): 201–207.

18. H. R. Varian, *Microeconomic Analysis* (Rockland, MA: R. S. Means, 1992).

19. K. K. Watson, J. H. Ghodasra, M. A. Furlong, and M. L. Platt, "Visual Preferences for Sex and Status in Female Rhesus Macaques," *Animal Cognition* 15 (2012): 401–407.

20. D. V. Smith, B. Y. Hayden, T.-K. Truong, A. W. Song, M. L. Platt, and S. A. Huettel, "Distinct Value Signals in Anterior and Posterior Ventromedial Prefrontal Cortex," *Journal of Neuroscience* 30 (2010): 2490–2495.

21. S. C. Stearns, "Daniel Bernoulli (1738): Evolution and Economics under Risk," *Journal of Biosciences* 25 (2000): 221–228; P. W. Glimcher, M. C. Dorris, and H. M. Bayer, "Physiological Utility Theory and the Neuroeconomics of Choice" *Games and Economic Behavior* 52 (2005): 213–256.

22. K. K. Watson and M. L. Platt, "Social Signals in Primate Orbitofrontal Cortex," *Current Biology* 22 (2012): 2268–2273; J. T. Klein and M. L. Platt,

"Social Information Signaling by Neurons in Primate Striatum," *Current Biology* 23 (2013): 691–696.

23. J. M. Pearson, K. K. Watson, and M. L. Platt, "Decision Making: The Neuro-ethological Turn," *Neuron* 82 (2014): 950–965.

24. J. T. Klein, S. V. Shepherd, and M. L. Platt, "Social Attention and the Brain," *Current Biology* 19 (2009): R958–R962.

25. C. Janiszewski, J. Chris, and W. Luk, "The Influence of Classical Conditioning Procedures on Subsequent Attention to the Conditioned Brand," *Journal of Consumer Research* 20 (1993): 171–189.

Beauty Matters in Ways We Know and in Ways We Don't

Anjan Chatterjee

WE ARE ALL DRAWN to visual beauty. We find ourselves attracted to beautiful people, places, and things.[1] The experience of beauty is vivid and immediate, and despite its being familiar, it harbors an ineffable mystery. Why should some configurations of line, shape, shadow, and color excite us so?

In this essay I focus on faces as an important example of objects that can be beautiful. Scientists are beginning to investigate what beauty means biologically. Fundamental to any scientific approach is measurement, an approach that some regard as anathema when applied to beauty. The measurement of facial beauty has an unfortunate history of parochial and racist agendas masquerading as claims to being objective and universal.[2] For example, the eighteenth-century Dutch artist and anatomist Petrus Campers developed facial profile measurements based on facial angles; one line of the angle was from the nostril to the ear and the other from the most protruding part of the jawbone to the most prominent part of the forehead. African and Asian angles were closer to 70 degrees and European angles closer to 80 degrees. He noted, "It is amusing to contemplate an arrangement . . . in regular succession: apes, orangs, negroes, the skull of a Hottentot, Madagascar, Celebese, Chinese, Mogulier, Calmuk and diverse Europeans."[3] This ordering was used to claim that European features were at the top of an objective hierarchy of beauty.[4]

Despite this murky history, many experiments demonstrate that facial beauty is associated with features that have nothing to do with a person's race or ethnicity. These features include averaging, symmetry, and the physical effects of estrogen and testosterone. Why do we find faces with these features beautiful? Two evolutionary forces, natural selection that enhances survival and sexual selection that enhances reproduction, offer insight into this question.[5]

In the laboratory, averaged features are constructed with computer programs that combine different faces. Although individual differences in preferences for faces certainly exist, averaged faces are typically regarded as more attractive than any individual face contributing to the composite. This effect is seen in Western, Chinese, and Japanese observers for both within and across race averages.[6] Averaged faces in the laboratory are analogous to faces that represent the central tendencies of populations in real life.[7] Presumably physical features that are a mix of different populations signal greater genetic diversity. This diversity also implies greater flexibility to adapt and survive under changing environmental conditions. Averaged features are thought to advertise the fitness of a person. These speculations apply to the last several million years during which our brains have evolved and may not apply in the current environment of modern medicine and technological innovation. Despite differences in our current environment, some studies suggest an association of averageness and health in perceived health and in medical records.[8]

The evolutionary argument for beauty in facial symmetry follows a similar logic of advertising fitness.[9] Many developmental abnormalities produce physical asymmetries. Symmetry also indicates a healthy immune system. Parasites, which played an important role in human evolution, produce physical asymmetries in many plants, animals, and humans. Humans differ in their susceptibility to parasites, based in part on the strength of their immune systems. So facial and bodily symmetry advertise a person who is resistant to parasites. Consistent with this idea, while attractiveness is highly regarded in general, data from twenty-nine cultures indicate that people in geographic areas with a greater prevalence of pathogens value a mate's physical attractiveness more than people with relatively little pathogen infestations.[10]

Sexual dimorphism is a third parameter that contributes to why many of us find certain physical features attractive.[11] Feminine features rendered by estrogen signal fertility. Men are attracted to women with facial features that advertise fertility by balancing youth and maturity. Faces that are too childlike might mean that a girl is not yet fertile; women need some degree of sexual maturity to bear and raise children. Thus men find women with big eyes, full lips, narrow chins (indicators of youth), and high cheekbones (indicators of sexual maturity) attractive. Younger women have a longer time to bear children than do older women. Men drawn to young, fertile women are likely to have more children than men drawn to older women, and this preference is passed on to subsequent generations.

Physical features that make male faces attractive can also be explained by evolutionary mechanisms. Testosterone produces masculine features with large, square jaws, thinner cheeks, and heavier brows. In many species, including humans, testosterone appears to suppress the immune system.[12] So the idea that features affected by testosterone are fitness indicators doesn't make sense. Here, the logic is turned on its head. Rather than a fitness indicator, scientists invoke a handicap principle.[13] Only men with strong immune systems can pay the price that testosterone levies on their immune systems. The most commonly cited example of a handicap is the peacock and its tail. Certainly this cumbersome but beautiful tail doesn't exactly help the peacock to approach peahens or avoid predators with agility. Why would such a handicapping appendage evolve? The reason is that the peacock is also advertising its strength to the peahen. It can afford to spend excess energy maintaining a costly tail. Consistent with this idea, the brightest birds are found in areas with most parasites, again suggesting that especially fit birds are resistant to such infections and show off by diverting resources to extravagant appendages.[14] With masculine facial features for which the immune system takes a hit, the argument is that the owner of such a face is so genetically fit that he can afford to spend fitness capital on his square jaw.

Of course, most young people are driven by desire when choosing a mate and not an analytic calculation about maximizing gene survival projected into an indefinite future. However, attraction to people who have the potential to produce more and healthier children is a trait that, over evolutionary time, has become an intrinsic feature of our brains.

How does the brain respond to facial beauty? To investigate something as complex as an aesthetic response, the phenomenon needs to be broken down into component parts, including sensory-motor, emotion-reward, and meaning-knowledge systems.[15] The visual system is organized such that different categories of visual objects, such as faces, landscapes, and written words, are processed in different areas. When people look at faces, parts of the visual cortex important for processing faces and objects are activated more strongly if the faces are attractive. This occurs in the fusiform gyrus, a part of the visual brain that is tuned to faces, and an adjacent area called the lateral occipital complex, which responds to objects. Beyond this response in visual processing areas, attractive faces engage reward systems. Reward systems are areas of the brain that respond to a variety of rewards, such as to food when one is hungry. These areas include parts of the ventral striatum, the orbitofrontal cortex, and the ventromedial prefrontal cortex. Thus the visual cortex tuned to processing faces interacts with reward systems in the brain to underpin the experience of beauty.[16]

It turns out that our brains respond to beautiful faces even when we are not thinking about beauty.[17] We conducted a study in which people lying in a brain-scanning machine judged the identity of a face—that is, whether one face was of the same person they saw in the preceding image. Parts of their visual cortex responded more robustly to attractive faces (as judged by the group of participants) even though they were answering questions about a person's identity and not his or her beauty.[18] Similarly, another research group asked people to judge a perceptual feature: the width of the face. They also found an automatic neural response to facial beauty, despite the fact that respondents were making a perceptual and not an aesthetic judgment. This neural response occurred in parts of the brain's reward circuits.[19] Taken together, these studies suggest that automatic beauty detectors in the brain link vision and pleasure. These detectors seem to ping every time when we see beauty, regardless of whatever else we might be thinking.

Most of us have had the experience that getting to know people can change how attractive we find them. If we like them, we may find them more attractive over time. This is an example of the meaning-knowledge system modulating aesthetic experiences—what we know about people can make them appear more or less beautiful. We have an inborn "beauty is good" stereotype.[20] In trying to understand the biology of these stereo-types, investigators have examined the neural responses to attractive faces and morally good acts. People were shown faces that were either beautiful or neutral and statements or pictures that depicted morally good or neu-tral acts. Parts of the brain that respond to rewards in the orbitofrontal cortex also respond to both facial beauty and moral correctness, suggest-ing that the reward experienced for beauty and goodness is similar in the brain.[21] It is interesting that a pattern of overlapping neural activity in the lateral orbitofrontal cortex is observed even when people are not thinking about beauty or goodness. Analogous to the automatic beauty detector, it is as if we automatically associate beauty with goodness, even when we are not thinking explicitly about either. This reflexive association may be the biological trigger for social effects of beauty that have been well doc-umented by social scientists—for example, attractive people receive many advantages in life, like higher pay and lesser punishments. These obser-vations reveal beauty's ugly hold on us. We recently found that people with minor facial anomalies are judged to be less good, competent, intel-ligent, hardworking, and kind. Unfortunately, most people also have a "disfigured-is-bad" stereotype. Understanding these cognitive biases built into our brains is critical to overcoming them if we wish to treat people fairly and judge them on the merits of their behavior and not the partic-ulars of their looks.

I leave you to ponder one final thought. The universal attributes of facial beauty were sculpted by environments that selected specific prefer-ence and trait combinations that were critical to health and reproductive success during the almost two million years of the Pleistocene epoch (which ended only about 11,700 years ago, a blink in evolutionary time). In modern societies, these same selection constraints no longer apply. For example, in much of the (technologically developed) world, parasite infestation is not a major cause of death. From antibiotics to surgeries, birth control to in vitro fertilization, the contemporary world that we have created is changing its selection criteria for reproductive success. Under these new conditions, combinations of physical traits and preferences that promote survival can change. As we profoundly alter our environ-ment, modern medicine and technological innovations are profoundly altering the very nature of beauty.

NOTES

1. A. Chatterjee, *The Aesthetic Brain: How We Evolved to Desire Beauty and Enjoy Art* (New York: Oxford University Press, 2014).

2. J. S. Haller, *Outcasts from Evolution: Scientific Attitudes of Racial Inferiority, 1859–1900* (Urbana: University of Illinois Press, 1971).

3. Quoted in M. Kemp, "Slanted Evidence," *Nature* 402 (1999): 727.

4. M. C. Meijer and P. Camper, "Petrus Camper on the Origin and Color of Blacks," *History of Anthropology Newsletter* 24 (1997): 3–9.

5. K. Grammer, B. Fink, A. P. Møller, and R. Thornhill, "Darwinian Aesthetics: Sexual Selection and the Biology of Beauty," *Biological Reviews* 78 (2003): 385–407; G. Rhodes, "The Evolutionary Psychology of Facial Beauty," *Annual Review of Psychology* 57 (2006): 199–226.

6. G. Rhodes, S. Yoshikawa, A. Clark, K. Lee, R. McKay, and S. Akamatsu, "Attractiveness of Facial Averageness and Symmetry in Non-Western Cultures: In Search of Biologically Based Standards of Beauty," *Perception* 30 (2001): 611–625.

7. While averaging faces smoothes out blemishes, deformities, and asymmetry, the question of blemishes has been addressed with modern averaging techniques. Some studies have also parceled out contributions of symmetry. How much variance in beauty ratings is explained by symmetry and by averaging is debated among people who study this issue.

8. G. Rhodes, L. A. Zebrowitz, A. Clark, S. M. Kalick, A. Hightower, and R. McKay, "Do Facial Averageness and Symmetry Signal Health?" *Evolution and Human Behavior* 22 (2001): 31–46.

9. R. Thornhill and S. W. Gangestad, "Facial Attractiveness," *Trends in Cognitive Sciences* 3 (1999): 452–460.

10. S. W. Gangestad and D. M. Buss, "Pathogen Prevalence and Human Mate Preferences," *Ethology and Sociobiology* 14 (1993): 89–96.

11. D. I. Perrett, K. J. Lee, I. Penton-Voak, et al., "Effects of Sexual Dimorphism on Facial Attractiveness," *Nature* 394 (1998): 884–887.

12. C. J. Grossman, "Interactions between the Gonadal Steroids and the Immune System," *Science* 227 (1985): 257–261; J. Alexander and W. H. Stimson, "Sex Hormones and the Course of Parasitic Infection," *Parasitology Today* 4 (1988): 189–193.

13. A. Zahavi, *The Handicap Principle: A Missing Piece of Darwin's Puzzle* (Oxford: Oxford University Press, 1997).

14. W. D. Hamilton and M. Zuk, "Heritable True Fitness and Bright Birds: A Role for Parasites," *Science* 218 (1982): 384–387.

15. A. Chatterjee and O. Vartanian, "Neuroaesthetics," *Trends in Cognitive Sciences* 18 (2014): 370–375.

16. J. Winston, J. O'Doherty, J. Kilner, D. Perrett, and R. Dolan, "Brain Systems for Assessing Facial Attractiveness," *Neuropsychologia* 45 (2007): 195–206.

17. J. Sui and C. H. Liu, "Can Beauty Be Ignored? Effects of Facial Attractiveness on Covert Attention," *Psychonomic Bulletin and Review* 16 (2009): 276–281.

18. A. Chatterjee, A. Thomas, S. E. Smith, and G. K. Aguirre, "The Neural Response to Facial Attractiveness," *Neuropsychology* 23 (2009): 135–143.
19. H. Kim, R. Adolphs, J. P. O'Doherty, and S. Shimojo, "Temporal Isolation of Neural Processes Underlying Face Preference Decisions," *Proceedings of the National Academy of Sciences* 104 (2007): 18253–18258.
20. K. Dion, E. Berscheid, and E. Walster, "What Is Beautiful Is Good," *Journal of Personality and Social Psychology* 24 (1972): 285–290.
21. T. Tsukiura and R. Cabeza, "Shared Brain Activity for Aesthetic and Moral Judgments: Implications for the Beauty-Is-Good Stereotype," *Social Cognitive and Affective Neuroscience* 6 (2011): 138–148.

"Man Can Do What He Wants, but He Cannot Will What He Wants"

Scott M. Sternson

IN 1839, Arthur Schopenhauer was a fifty-two-year-old German philosopher living as an isolated private scholar in Frankfurt. He had notably high self-regard, but his contemporaries did not share this notion. Schopenhauer was also deeply pessimistic about human nature, with one result being that he never married. (On marriage: "Marrying means, to grasp blindfold into a sack hoping to find an eel out of an assembly of snakes.") He was not influential for the first fifty years of his life, but inherited wealth allowed him to pursue his scholarly interests.

Seeking recognition, Schopenhauer entered an essay contest held by the Royal Norwegian Society of Sciences that asked, "Can the freedom of the will be proven from self-consciousness?" It is crucial that Schopenhauer's lack of academic status was not a serious impediment for competing because the essays were submitted anonymously. His essay defined "the will" as the underlying motivation that guides choices. Schopenhauer systematically argued that although individuals were free in their actions, the will was not free—it was constrained by inborn factors. Nearly one hundred years later, Albert Einstein neatly paraphrased Schopenhauer's essay: "Man can do what he wants, but he cannot will what he wants." With his essay *against* the freedom of the will, Schopenhauer won the prize, and with the resulting attention that it brought, he went on to become a major influence on late nineteenth- and twentieth-century thinkers.[1]

Schopenhauer's point was that we do not really know why we do things, and there is no reason to believe that we are consciously willing these motivations. In fact, our will is generated subconsciously, and our self-consciousness figures out what to do with these needs and desires. Because our brain controls our behavior, the origin of the will is ultimately a question for neuroscience. Moreover, our brain has evolved over hundreds of millions of years. Therefore, the source of our motivations is deeply intertwined with evolutionary biology, specifically in the way that natural selection forms neural mechanisms that compel organisms to perform behaviors essential for survival.

Why we do something is a product of three different processes: innate sensory hedonics (i.e., preferences), learning, and instinct. Sensory hedonics refers to the fact that certain stimuli, such as sweet tastes, are innately pleasurable and bitter tastes are not, as evidenced by studies in infants immediately after birth.[2] These innately pleasurable or undesirable stimuli can also drive reinforcement learning. Positive reinforcement is the process of learning to perform actions that increase the receipt of a pleasurable outcome, and negative reinforcement involves learning to behave in a way to avoid unpleasant outcomes. Instinct plays a role where complex behaviors seem to be produced without learning. Caution is required, though, because instinct is also a frequently underscrutinized term, and it can be difficult to tell which aspects of behavior are instinctive and which are learned.

To illustrate this, let's consider feeding behavior in the laughing gull, a bird commonly found on the eastern coast of the United States. Like many bird families, the laughing gull regurgitates food into the mouth of its newborn offspring. The laughing gull chick must peck and pull at the parent's beak until it regurgitates food that can be consumed by the chick. It is easy to conclude that the newborn bird possesses a feeding instinct that consists of a complex series of behaviors.

However, in the 1960s, a series of studies by Professor Jack Hailman systematically examined this "feeding instinct" in newborn laughing gulls. (The work is summarized in an article cheekily titled "How an Instinct Is Learned.")[3] Using accurate models of adult laughing gull heads whose design Hailman could manipulate, he analyzed which particular features induced pecking by the newborn chicks. It is interesting that the presence of traces of food on the models did not increase pecking. This indicated that, at least initially, pecking may not be a food-seeking behavior. Instead, Hailman used models of increasing simplicity that ultimately revealed that newborn birds would peck at a simple wooden dowel, preferring those with a length and diameter that best mimicked the beak of an adult laughing gull. Moreover, pecking was greatest when the dowel moved at a similar speed as that of the adult bird. Instead of an instinct to perform a series of behaviors directed specifically toward food, it is a reflex for the laughing gull to peck at a cylinder with specific diameter and speed. Reflexes are unlearned simple movements that happen reliably in response to particular stimuli. This is just the type of simple stimulus-response relationship that natural selection seems to hard-wire into brains. Another example may be seen in the organization of the eye's retina. Similar visually guided prey capture behaviors are observed in frogs, salamanders, and dragonflies that are tuned for optimal-sized objects moving with optimal speed (see also the essay by Aniruddha Das in this volume).[4] In humans, related reflexes include the rooting and sucking responses that are initially elicited by nearly any contact with an infant's cheek or tongue but that become productive for feeding when the contact is with a lactating breast.

Food seeking in the laughing gull chick is a survival behavior that is comprised of the basic reflex to peck at a cylindrical object and is linked to the fact that the main source of moving cylindrical objects in the newborn bird's environment is a parent's beak. This is further coupled with the fact that beak pecking in the adult bird leads to food regurgitation, making it available for consumption by the fledgling. The process is a simple mechanical relationship that, initially, requires no behavior that is directed toward food. This is where learning comes into play. Performing these reflexes and, as a result, receiving food, *shapes* beak pecking into food-seeking behavior. This is the role of reinforcement learning, in which a desirable outcome leads to a specific set of neurochemical processes that make the actions that led to that outcome more likely to occur in the future. In this way, the bird's brain is wired with a spring-loaded trap combining two very simple innate behaviors (newborn pecking and parental regurgitation) such that, once the trap is sprung, feeding behavior is captured through reinforcement learning. This "behavior trap" is such a simple and robust process that acquiring this feeding behavior is essentially foolproof and appears unlearned, even though it is not.

With this example in mind, consider the origins of Schopenhauer's "will" and how it underlies our behaviors. Eating, aggression, parenting, sexual preference, or social interactions are extremely common and complex behaviors that are critical for survival of either the organism or its species. But where do these motivations come from? What is learned and what is innate? In most cases this is not well established in people, but studies from other species give some intriguing clues.

In the field of neuroscience, several remarkable discoveries have shown that artificial activation of specific brain regions with electrodes can elicit various survival behaviors when they normally wouldn't occur. The part of the brain that seems to play an outsized role in instinctive motivations is a hazelnut-sized structure at the bottom of our brains called the hypothalamus. For instance, stimulation of one part of the hypothalamus promotes eating, another elicits drinking, and yet another leads to fighting.[5] Thus there are specialized circuits in the brain that mediate these evoked behaviors. These circuits are substrates by which natural selection can influence complex behaviors through simple behavior traps that combine innate sensory hedonics, learning, and motivational need states. The latter state involves the sense of discomfort that we feel when a basic component necessary for a physical function is lacking, such as the unpleasant need for food in a hungry animal. The core circuits that mediate these different elements of our preferences are the origin of our will, and they ensure that, like the laughing gull, we will engage in essential behaviors.

We have some insight into how this occurs for the most elemental motivations, such as hunger. When an animal has not eaten for a while, specific hypothalamic brain cells called AGRP neurons become activated, and the activation promotes feeding. AGRP neurons are then inhibited as soon as food is detected and eaten.[6] Although these neurons are closely associated with eating, the activity of these cells is experienced as unpleasant, indicating that animals appear to eat, at least in part, to shut these neurons off.[7] Thus this system "pushes" a mouse to engage with the environment to find food in order to avoid an unpleasant hunger state. AGRP neurons are complemented by other circuits that ensure that certain experiences, such as sweet taste, are also innately pleasurable, thereby providing an incentive to continue eating. These two circuits comprise a push-pull system for learning behaviors that results in eating by the avoidance of an unpleasant hunger state and by a reflexive tendency to engage in consumption of pleasant tastes. Because both of these processes drive learning, they are core constituents of a spring-loaded trap for behaviors that cause an animal to obtain and ingest food.

The relationship between instinctual motivations and learning is at the heart of many complex behaviors. For example, parenting behavior in mice is profoundly influenced by anterior portions of the hypothalamus. Stimulation of neurons in this region can make male mice, otherwise prone to infanticide, behave as caring parents. Conversely, silencing the electrical output of this brain region can turn otherwise good fathers and mothers into child killers.[8] In another example of complex behavior, the elimination of a specialized chemical sensing zone in the nose in female mice will transform their sexual behavior to resemble that of males. They chase females and males alike and attempt to mount them.[9] Activation of other hypothalamic regions can also directly evoke sexual activity.[10] These examples illustrate how sexual and parenting behaviors are not fixed but rather are shaped by the interaction of external cues and internal processes that result in specific sex roles.

Because our hypothalamus is similar to that of other animals, human nature is also likely to be derived from evolutionarily conserved spring-loaded traps wired in our neural circuitry. These circuits bias our responses to sensory experiences to produce behaviors that deal with needs and desires that are encoded by dedicated brain circuits. What makes us uniquely human are the elaborate lengths to which we develop behaviors that satisfy these needs. The richness of human behavior results from our big brains responding in diverse ways to our innate needs and desires. Human happiness as well as dissatisfaction is tied to satisfying basic innate motivations.

These innate motivations do not fully form our behaviors. Instead, instinct provides a predisposition that guides learning about how to cope with intrinsic motivational processes that are experienced as social and sexual attraction, anxiety, fear, and physiological needs such as hunger and thirst. Feelings of dissatisfaction, such as stress, can originate from the requirement to balance competing needs—a familiar challenge of life.

Perhaps most important for humans, we are compelled by basic neurological processes to be social. Sociality is imposed on behavior by natural selection and is thus part of brain structure. This has been demonstrated by showing that monogamous and polygamous vole species differ by the absence of a specific receptor in the brain of the polygamous vole. Moreover, forcing expression of the receptor transforms the polygamous species into a monogamous one.[11] More generally, social animals find it necessary to constrain expression of some motivations that are related solely to the needs and desires of the individual in order to get along with a group. Coupled with human intellect and our capacity for analytical self-reflection, the benefits and constraints of social life lead to laws and rules for interacting with each other (Schopenhauer: "Compassion is the basis for morality.")[12] Of course, the precise manner in which this is carried out is often controversial among different groups, and competing approaches to balancing individual and social actions are played out in both philosophy and, in more explicit forms, politics.

Neuroscience has reached a stage where further insight into the basis of the human experience is within reach through the careful study of the biology of our motivations. Ultimately, a deeper understanding of the neurologically derived will is essential to understanding ourselves, including the drivers of our well-being and even the structures of our societies.

NOTES

1. D. E. Cartwright, *Schopenhauer: A Biography* (New York: Cambridge University Press, 2010). For Einstein quote, see http://www.einstein-website.de/z_biography/credo.html.

2. K. C. Berridge, "Measuring Hedonic Impact in Animals and Infants: Microstructure of Affective Taste Reactivity Patterns," *Neuroscience and Biobehavioral Reviews* 24 (2000): 173–198.

3. J. P. Hailman, "How an Instinct Is Learned," *Scientific American* 221 (1969): 98–106.

4. J. Y. Lettvin, H. R. Maturana, W. S. McCulloch, and W. H. Pitts, "What the Frog's Eye Tells the Frog's Brain," *Proceedings of the Institute of Radio Engineers* 47 (1959): 1940–1951; D. O'Carroll, "Feature-Detecting Neurons in Dragonflies," *Nature* 362 (1993): 541–543. The spacing of cells in the retina and the manner in which they are connected influences the size, speed, and direction of the objects that they will optimally detect. Thus the organization of the retina can be one site on which natural selection acts to determine the size of an object that, for example, a frog will detect as prey or the fact that a frog will capture only moving prey.

5. S. M. Sternson, "Hypothalamic Survival Circuits: Blueprints for Purposive Behaviors," *Neuron* 77 (2013): 810–824.

6. J. N. Betley, S. Xu, Z. F. Cao, R. Gong, C. J. Magnus, Y. Yu, and S. M. Sternson, "Neurons for Hunger and Thirst Transmit a Negative-Valence Teaching Signal," *Nature* 521 (2015): 180–185; Y. Chen, Y. C. Lin, T. W. Kuo, and Z. A. Knight, "Sensory Detection of Food Rapidly Modulates Arcuate Feeding Circuits," *Cell* 160 (2015): 829–841.

7. External activation of AGRP neurons by an experimenter leads to avid eating but also to avoidance of places and flavors consumed when the neurons were active, indicating that these neurons mediate some of the negative feelings associated with hunger.

8. Z. Wu, A. E. Autry, J. F. Bergan, M. Watabe-Uchida, and C. G. Dulac, "Galanin Neurons in the Medial Preoptic Area Govern Parental Behaviour," *Nature* 509 (2014): 325–330.

9. T. Kimchi, J. Xu, and C. Dulac, "A Functional Circuit Underlying Male Sexual Behaviour in the Female Mouse Brain," *Nature* 448 (2007): 1009–1014.

10. H. Lee, D. W. Kim, R. Remedios, T. E. Anthony, A. Chang, L. Medisen, H. Zeng, and D. J. Anderson, "Scalable Control of Mounting and Attack by Esr1+ Neurons in the Ventromedial Hypothalamus," *Nature* 509 (2014): 627–632.

11. M. M. Lim, Z. Wang, D. E. Olazabal, X. Ren, E. F. Terwilliger, and L. J. Young, "Enhanced Partner Preference in a Promiscuous Species by Manipulating the Expression of a Single Gene," *Nature* 429 (2004): 754–757.

12. Quoted in Cartwright, *Schopenhauer*, p. 167.

The Brain Is Overrated

Asif A. Ghazanfar

Alcor's interest is preserving people. In the entire human body, there is one organ that is absolutely essential to personhood: *the brain*. Injuries outside the brain are wounds to be healed. Injuries to the brain are injuries to *who we are*.

—*Alcor Life Extension Foundation website (2015)*

We used to think that knowing our genomes would tell us much of what we needed to know about ourselves. ("Today we are learning the language in which God created life," President Bill Clinton stated at the Human Genome Project celebration in 2000). Recently we've been shifting our thinking toward another idea: we are our *connectomes*—who we are as individuals is due simply to the specificity and totality of connections and other properties of all the neurons of our nervous system. This idea stems from the belief that the brain is like a control center for behavior. It plans future actions based on information it receives from the environment and then generates commands for movements based on those plans. If we could map and measure every one of the roughly 100 trillion connections that our 100 billion neurons collectively make, could we understand the mind and our individual differences? Could we live forever by uploading this knowledge into a computer so that it in effect ran the software of our brain? Or could we do so by reanimating a cryogenically preserved brain and putting it into another, younger body? Does the brain really contain *everything* important about behavior and who we are as individuals? No, it does not.

Your behavior is, for sure, dependent on your individual neural circuits, but it is just as dependent upon your individual body. Different parts of the body act as filters for both incoming and outgoing signals.[1]

Your ears, for example, are very important for identifying where a sound is coming from. That's obvious. What's not so obvious is that locating whether that sound is coming from above or below you is dependent upon the shape of your ear. The ridges and valleys of the outer ear filter sounds—making some parts of the sound louder and others softer—before they hit the eardrum. It is critical that the parts of a given sound that get louder or softer also depend on whether the sound is hitting the outer ear from above or below, and we thus learn to associate those acoustic differences with the location of the sound source. You may have noticed that ears are a little like fingerprints and that everyone has a unique set (indeed, biometric research has begun using ears as unique identifiers).[2] So how you locate sounds is dependent upon the exact shape of your particular ears. Your brain, in effect, learns about your ear shape through experience. To put it another away, if you changed the shape of your ears by, say, attaching pieces of Silly Putty to them, you would degrade your ability to locate a sound.

Your body also filters outgoing signals that drive your muscles to produce movements. A good example is your voice. Your voice is distinctive, unique. All those who know you well can identify you easily just by listening to you. Their recognition may be partly due to your stereotypical word choices and manner of speaking, but it is also due to the unique shape of your oral and nasal cavities. Voice production involves a sound source—produced by respiration driving the movements of the vocal cords—coupled to a sound filter consisting of the oral and nasal cavities above the vocal cords. The vocal cords generate a sound that is a bit like a buzzing and whose pitch depends upon how fast the vocal cords are moving. This buzzing sound then passes through the vocal tract that filters it. That is, like what our outer ears do to sounds going in, the shape of the vocal tract causes some parts of the sound to get louder and others to get quieter as they go out. We produce different vowel sounds by changing the shape of our vocal tract with different facial expressions (for example, producing an /a/ sound versus an /i/ sound). However, some parts of the vocal tract are fixed and individually distinctive, representing individual variation in how the oral and nasal cavities developed. These unique features filter the sound in such a way as to give us our voice identity. Both how we hear and how we speak are tied to the shape of our bodies.

The importance of our bodies to behavior and experience is reflected in how it changes and guides the nervous system during development. For example, we are able to localize sound pretty well at a very young age, but our bodies are still growing—our ears are still changing their shape. So even though those filtering properties of the ears are changing, the brain is constantly recalibrating itself to account for these changes. In fact, the auditory system is so dependent upon the shape of the ears to guide its function that in order to properly function, it has to wait for the body to catch up to it. If you record the activity from auditory neurons in a very young animal while it is listening to sounds, you will find that the neurons don't function very well in terms of being able to determine where a sound is coming from. Typically, it is assumed that this must be because the nervous system is still immature. However, if you artificially give this same young animal the ears of an adult (you can do this by using virtual reality, putting sounds directly in the animal's ear canals after they've been filtered by a simulated adult ear), then quite suddenly those neurons function perfectly well and encode sound location accurately. Thus the body is guiding the function of the nervous system, not the other way around.

Here's another, perhaps more memorable, example demonstrating how the body shapes the nervous system. Male mammals have a specialized region in their spinal cords that is involved in urination and penile erection (called "Onuf's nucleus" in humans; a "nucleus" in the brain is simply a group of densely packed neurons); the same region is present in females but is much smaller in size. The size difference is due to the differing shape of the male versus female body. This is obvious, of course, given that part of Onuf's nucleus function is for erections, but the developmental mechanism for how the difference comes about is quite elegant.[3] Early in development *both* males and females start off having the muscles that could putatively control erections and urination, the bulbocavernosus and levator ani. Both muscles have testosterone receptors on them, but only male infants produce lots of testosterone during early development. The survival of the muscle is dependent upon testosterone binding to those receptors. The paucity of testosterone in females causes the muscles to degenerate; the bulbocavernosus disappears, and the levator ani is reduced in size. The nervous system, in turn, uses a clever matching process to make sure it has all the neurons it needs, no more and no fewer. For their survival, neurons need molecular nutrients ("trophic factors") delivered by the muscles; the bigger the muscle, the more trophic factors are available and thus more neurons can survive. When the bulbocavernosus and levator ani muscles in females atrophy due to lack of testosterone, they are also reducing the amount of trophic factors, which then causes Onuf's nucleus to shrink in size as many of the neurons die off. This is just one example of how the body shapes the developing nervous system.

The body not only filters incoming and outgoing signals and sculpts the developing nervous system, but its material properties can also make the brain's job easier.[4] Animals often produce many different types of calls to communicate. A standard assumption by scientists would be that each different type of call is produced by a call-specific neural activity pattern. It turns out, however, that parts of the vocal apparatus (the vocal cords and the vocal tract) interact in ways that are nonlinear (or chaotic), depending on how much respiratory power is put through the system and/or how much tension there is in the vocal cords. These nonlinearities generate very distinct sounds without requiring any sophisticated neural activity patterns. If you've ever tried to sing too high a note, then you are familiar with at least one type of vocal nonlinearity—the breaking of your voice. Another example comes from vocal development. In some monkey species, infants babble like human babies and produce long sequences of different types of vocal sounds.[5] Each type of vocalization is not the result of a special, dedicated pattern of neural activity but rather a single pattern of neural activity that simply goes up and down. This cyclical neural activity drives the respiratory power up and down. As it does so, it is the nonlinearities of the vocal apparatus that generate the different sounds. Specific behaviors need not emanate from the nervous system via its specific activity patterns; the material properties of the body also help generate behavioral diversity.

You are not you without your brain *and* your body and their integrated experiences. Your body is as important as your brain when it comes to behavior. The body acts as a filter for what is experienced and helps determine identity. The changing body shape guides how the nervous system develops. Finally, the material properties of the body can make the operations of the nervous system simpler (generating different vocal sounds). Thus in many important ways, the functions of the brain do not adhere to what can be seen in their connections or activity patterns. "It is the man, not the brain, that thinks; it is the organism as a whole and not one organ that feels and acts," wrote the philosopher George H. Lewes in 1891.[6] He was correct not only in spirit but in logic. It is a fallacy (the mereological fallacy, to be specific) to ascribe behavioral attributes to parts of an organism that can only intelligibly be ascribed to the animal as a whole.

So let's say you shelled out approximately $80,000 to become a "member" of a life-extension foundation and had your brain cryopreserved. If they find a younger body and place your brain into it, then your brain could not wire up accurately to that body because your brain circuits were shaped by the development and experience linked to your original body. Moreover, you would not look the same, sound the same, or feel the same as before to yourself or anyone who knew you before. But let's allow for two additional factors: you don't care about being a different external self, and the regenerative technology is such that the old brain can rewire itself to the new body. If the latter is true and your brain changes according to the shape of this new body and the new experiences it generates, then your history—which is intimately linked to your body—would be erased. You would then be somebody else.

NOTES

1. H. J. Chiel and R. D. Beer, "The Brain Has a Body: Adaptive Behavior Emerges from Interactions of Nervous System, Body and Environment," *Trends in Neurosciences* 20 (1997): 553–557.

2. A. H. Cummings, M. S. Nixon, and J. N. Carter, "A Novel Ray Analogy for Enrolment of Ear Biometrics," Fourth IEEE International Conference: Biometrics: Theory Applications and Systems, Washington, D.C., September 27–29, 2010, pp. 1–6.

3. A. Matsumoto, P. E. Micevych, and A. P. Arnold, "Androgen Regulates Synaptic Input to Motoneurons of the Adult Rat Spinal Cord," *Journal of Neuroscience* 8 (1988): 4168–4176.

4. R. Pfeifer and J. Bongard, *How the Body Shapes the Way We Think: A New View of Intelligence* (Cambridge, MA: MIT Press, 2006).

5. Y. S. Zhang and A. A. Ghazanfar, "Perinatally Influenced Autonomic System Fluctuations Drive Infant Vocal Sequences," *Current Biology* 26 (2016): 1249–1260.

6. G. H. Lewes, *The Physical Basis of Mind* (Boston: Houghton Mifflin, 1891), p. 498.

Dopamine Made You Do It

Terrence Sejnowski

DOPAMINE-RELEASING NEURONS are a core brain system that controls motivation.[1] When enough dopamine-releasing neurons die, the symptoms of Parkinson's disease appear; these include motor tremor, difficulty initiating actions, and, eventually, anhedonia, the complete loss of pleasure in any activity. The end stage includes catatonia, a complete lack of movement and responsiveness. But when the dopamine neurons are behaving normally, they provide brief bursts of dopamine to the neocortex and other brain areas when an unexpected pleasure (reward) occurs (be it food, money, social approval, or a number of other things) and a diminution of activity when less than expected reward is experienced (this can be a smaller reward or no reward at all).

Your dopamine neurons can be polled when you need to make a decision. What should I order from the menu? You imagine each item, and your dopamine cells provide an estimate of the expected reward. Should I marry this person? Your dopamine cells will give you a gut opinion that is more trustworthy than reasoning. Problems with many different dimensions are the most difficult to decide. How do you trade off a sense of humor in a mate, a good dimension, against being messy, a bad dimension, or hundreds of other comparisons? Your brain's reward systems reduce all these dimensions down to a common currency, the transient dopamine signal. Dopamine neurons receive inputs from a part of the brain called the basal ganglia, which in turn receive input from the entire cerebral cortex. The basal ganglia evaluate cortical states and are involved with learning sequences of motor actions to achieve a goal.

The dark side of reward is that all addictive drugs act by increasing the level of dopamine activity. In essence, drugs like cocaine and heroin (as well as nicotine and alcohol) hijack the dopamine reward system, making your brain believe that taking the drug is your most important immediate goal. Withdrawal symptoms dominate when drugs are not immediately available. This motivates desperate actions to obtain more drugs, and such actions can jeopardize life and livelihood. Even after an arduous rehabilitation process, which can take years, the brain's reward circuit is still altered by the experience of addiction, leaving the recovering addict vulnerable to a relapse. This can be triggered by people and places, sounds and smells previously associated with the drugs, or even paraphernalia used to take the drugs. For an addict, dopamine is deeply compelling.

The basal ganglia are part of all vertebrate brains. Within the basal ganglia the dopamine neurons mediate a form of learning called associative learning, made famous by Pavlov's dog. In Pavlov's experiment, a sensory stimulus such as a bell (a conditioned stimulus) was followed by the presentation of food (an unconditioned stimulus), which elicited salivation even without the bell (an unconditioned response). After several pairings, the bell by itself would lead to salivation (a conditioned response). Different species have different preferred stimuli to associate. Bees are very good at associating the smell, color, and shape of a flower with the rewarding nectar, and they use this learned association to find similar flowers that are in season. Something about this universal form of learning must be important, and there was a period in the 1960s when psychologists intensively studied the conditions that gave rise to associative learning and developed models to explain it.

Only the stimulus that occurs just before the reward becomes associated with the reward.[2] This makes sense since the stimulus is more likely to have caused the reward if it comes before the reward than a stimulus just after the reward. Causality is an important principle in nature.

Suppose you have to make a series of decisions to reach a goal. If you don't have all the information about the outcomes of the choices ahead of time, you have to learn as you make the choices in real time. When you get a reward after a sequence of decisions, how do you know which of the several choices you made were responsible? A learning algorithm that can resolve this issue, called the temporal credit assignment problem, was discovered by Richard Sutton at the University of Massachusetts at Amherst in 1988.[3] He had been working closely with Andrew Barto, his thesis adviser, on difficult problems in reinforcement learning, a branch of machine learning inspired by associative learning in animals. In temporal difference learning, you compare your expected reward for making a particular choice with the actual reward you get and change your expected reward so that next time you will be able make a better decision. Then an update is made to the value network that computes the future expected reward for each decision at each choice point. The temporal difference algorithm converges to the optimal series of decisions after you have had enough time to explore the possibilities. This is followed by a period of exploiting the best strategy found during the exploration.

Bees are champion learners in the insect world. It takes only a few visits to a rewarding flower for a bee to remember the flower. This fast learning was being studied in the laboratory of Randolph Menzel in Berlin when I visited him in 1992. The bee brain has around a million tiny neurons, and it is very difficult to record their electrical signals because they are so tiny. Martin Hammer in Menzel's group had discovered a unique neuron, called VUMmx1, that responded to sucrose (a type of sugar) with electrical activity but not to an odor; however, after the odor was delivered, followed shortly by the sucrose reward, VUMmx1 would now respond to the odor.

When I returned to La Jolla, Peter Dayan, a postdoctoral fellow in my lab who was an expert on reinforcement learning, immediately realized that this neuron could be used to implement temporal difference learning. Our model of bee learning could explain some subtle aspects of bee psychology, such as risk aversion. For example, when a bee is given a choice between a constant reward and twice the amount but at 50 percent probability (on average the same amount), bees prefer the constant reward. Read Montague, another postdoctoral fellow in my lab, took the next leap and realized that dopamine neurons in the vertebrate reward system may have a similar role in our brains.[4] In one of the most exciting scientific periods of my life, these models and their predictions were published and subsequently confirmed in monkeys with single neuron recordings by Wolfram Schultz and in humans with brain imaging.[5] Transient changes in the activity of dopamine neurons signal reward prediction error.

Temporal difference learning might seem weakly effective since the only feedback present is whether or not you are rewarded at the end of a sequence of actions. However, several applications of temporal difference learning have shown that it can be powerful when coupled with other learning algorithms. Gerry Tesauro worked with me on the problem of teaching a neural network to play backgammon. Backgammon is a highly popular game in the Middle East, and some make a living playing high-stakes games. It is a race to the finish between two players, with pieces that move forward based on each roll of the dice, passing through each other on the way. Unlike chess, which is deterministic, the uncertainty with every roll of the dice makes it more difficult to predict the outcome of a particular move. The knowledge of backgammon in Gerry's program was captured by a value function that provided an estimate of winning the match from all possible board positions as ranked by a panel of backgammon experts. A good move can be found simply by evaluating all possible moves from the current position and choosing the one with the highest value.

Our approach used expert supervision to train neural networks to evaluate game positions and possible moves. The flaw in this approach is that many expert evaluations of board positions were needed and the program could never get better than our experts. When Gerry moved to the IBM Thomas J. Watson Research Center, he switched from supervised learning to temporal difference learning and had his backgammon program play itself. The problem with self-play is that the only learning signal is a win or a loss at the end of the game with no information about the contribution of the many individual intermediate moves during the game to that win or loss.

At the beginning of the backgammon learning, the machine's moves were random, but eventually one side won. The reward first taught the program how to "bear off" and exit all of the pieces from the board at the end of the game. Once the endgame was learned, the value function for bearing off in turn trained the value function for the crucial middle game, where subtle decisions need to be made about engagements with the other player's pieces. Finally, after playing a hundred thousand games, the value function was honed to play the opening, in which pieces take defensive positions to prevent the other player from moving forward. Learning proceeds from the end of the game, where there is an explicit reward, back toward the beginning of the game, using the implicit reward learned by the value function. What this shows is that by back-chaining with a value function, it is possible for a weak learning signal like the dopamine reward system to learn a sequence of decisions to achieve a long-term goal.

Tesauro's program, called TD-Gammon, surprised me and many others when he revealed it to the world in 1992.[6] The value function had a few hundred model neurons in it, a relatively small neural network by today's standards. After a hundred thousand games, the program was beating Gerry, so he alerted Bill Robertie, an expert on positional play in backgammon from New York City, who visited IBM to play TD-Gammon. Robertie won the majority of games but was surprised to lose several well-played games and declared it the best backgammon program he had ever played. Several of the moves were unusual ones that he had never seen before; on closer examination these proved to be improvements on typical human play. Robertie returned when the program had reached a million self-played games and was astonished when TD-Gammon played him to a draw. A million may seem like a lot, but keep in mind that after a million games, the program saw only an infinitesimal fraction of all possible board positions. Thus TD-Gammon was required to generalize to new board positions on almost every move.

In March 2016, Lee Sedol, the Korean Go World Champion, played a match with AlphaGo, a program that learned how to play Go using temporal difference learning.[7] AlphaGo used neural networks with a much larger value network, having millions of units to evaluate board positions and possible moves. Go is to chess in difficulty as chess is to checkers. Even Deep Mind, the company that had developed AlphaGo, did not know its strength. AlphaGo had played hundreds of millions of games with itself, and there was no way to benchmark how good it was. It came as a shock to many when AlphaGo won the first three games of the match, exhibiting an unexpectedly high level of play. Some of the moves made by AlphaGo were revolutionary. AlphaGo far exceeded what I and many others thought was possible. The convergence between biological intelligence and artificial intelligence is accelerating, and we can expect even more surprises ahead. The lesson we have learned is that nature is more clever than we are.

We are just beginning to appreciate the powerful impact of dopamine on making decisions and guiding our lives. Since the influence of dopamine is subconscious, the story we tell ourselves to explain a decision is probably based on experiences no longer remembered. We make up stories because we need to have conscious explanations. Every once in a while we have a "gut feeling" about a choice that does not have an easy explanation—it was the dopamine that made us do it.

NOTES

1. E. Bromberg-Martin, M. Matsumoto, O. Hikosaka, "Dopamine in Motivational Control: Rewarding, Aversive, and Alerting," *Neuron* 68 (2010): 815–834.
2. There are some notable exceptions to the notion that the conditioned stimulus must immediately precede the unconditioned stimulus in associative learning. One is food-aversion learning. If you eat something and then become ill hours later, you will still strongly associate that food with the illness and tend to avoid that food in the future, even though the conditioned stimuli (the sight, smell, and taste of the food) can precede the unconditioned stimulus (feeling ill) by several hours.
3. R. S. Sutton, "Learning to Predict by the Method of Temporal Differences," *Machine Learning* 3 (1988): 9–44.
4. P. R. Montague, P. Dayan, and T. J. Sejnowski, "A Framework for Mesencephalic Dopamine Systems Based on Predictive Hebbian Learning," *Journal of Neuroscience* 16 (1996): 1936–1947.
5. W. Schultz, P. Dayan, and P. R. Montague, "A Neural Substrate of Prediction and Reward," *Science* 275 (1997): 1593–1599.
6. G. Tesauro, "Temporal Difference Learning and TD-Gammon," *Communications of the ACM* 38 (1995): 58–68.
7. D. Silver, A. Huang, C. J. Maddison, A. Guez, L. Sifre, G. van den Driessche, J. Schrittwieser, I. Antonoglou, V. Panneershelvam, M. Lanctot, S. Dieleman, D. Grewe, J. Nham, N. Kalchbrenner, I. Sutskever, T. Lillicrap, M. Leach, K. Kavukcuoglu, T. Graepel, and D. Hassabis, "Mastering the Game of Go with Deep Neural Networks and Tree Search," *Nature* 529 (2016): 484–489.

The Human Brain, the True Creator of Everything, Cannot Be Simulated by Any Turing Machine

Miguel A. L. Nicolelis

IT SEEMS THAT everywhere you look these days, whether in academia or in society at large, there is always someone propagating aloud the fallacious and misleading notion that the human brain is nothing but a glorified "meat" version of an otherwise typical digital machine, such as the computers we all use.[1] Hidden inside this assumption is the notion that, as just another digital device, the human brain, with all its unique functions, is simply an information processing device; not only could it one day be simulated and/or reproduced by a sophisticated digital computer, but also, at the limit, all its content, representing the entirety of each of our conscious and unconscious experiences over a lifetime, could be downloaded into a particular type of digital media. Following the same notion, complex content could be uploaded to one's brain so that, all of a sudden, one could become proficient in a new language or in a new subject.

This vision holds that to recreate or perpetuate one's existence, the only requirement needed is a new technology capable of extracting the digital information contained in one's brain. The key assumption behind this view is that the classic definition of information, introduced by Claude Shannon, can be applied in a straightforward manner to define the type of "currency" that brains like ours exchange routinely as they go about their business.[2] Yet even Shannon himself was well aware, back in the

1940s, that his definition of information—introduced in the context of quantifying message transmission through noisy communication channels, like the phone lines of his employer at the time, the Bell phone company—was not comprehensive enough to account for the semantically rich and meaning-imprinting nature of the messages produced by the human brain. In fact, more neurobiologically amenable definitions of information, which took into account the receiver's interpretation of the embedded meaning of messages generated by a transmitting source, were on the table in the early 1940s by the time Shannon came up with his own.[3] The shift from analog to digital computers that took place at the time, however, all but assured that Shannon's definition would take precedence because it could be easily implemented in the by then newly introduced digital circuits.[4] That did not mean, however, that the limitations of Shannon's definition in accounting for neurophysiological processes magically disappeared. Actually, they remain the same as they were in the 1940s.[5]

The core of the argument of what information really is, from a neurobiological point of view, has to do with the thesis defended by many authors that the human brain far exceeds the reach of digital computers.[6] Modern digital computers are the offspring of a more general computational concept, originally introduced by the British mathematician Alan Turing, and now classically known as the universal Turing machine.[7] In his classic paper in 1936, Turing introduced the theoretical computational framework of modern digital computing by demonstrating that if a given phenomenon can be reduced to a mathematical algorithm, it can be simulated by a universal Turing machine, the theoretical prototype of all digital computers ever created.[8] Turing's ingenious theoretical breakthrough was further expanded by the now classic Church-Turing assumption that any mathematical function that would be naturally considered as "computable" could be computed by a universal Turing machine.[9] It turns out, however, that nature is dominated by noncomputable phenomena, which, by definition, cannot be computed by a universal Turing machine. Thus most of what brains like ours do falls into the category of noncomputable phenomena. This historical insight illustrates that those who maintain that the human brain is just another digital system have not thought deeply enough about the intricacies of the neurophysiological mechanisms that allow sophisticated nervous systems like ours to do much more than simply carry out some sort of information processing.

When this absurd proposition remained confined to the domain of Hollywood science fiction movies, it did not matter much. Yet as soon as some computer scientists, and even neuroscientists, began to repeat this mantra in public and request billions of dollars from the European and U.S. taxpayers to pursue meaningless attempts to emulate the human brain in digital media, matters started to sound much more troublesome. That is why the moment I received the kind invitation from the editor of this volume to write a short essay about "the thing I most wanted everyone to know about the human brain," I did not hesitate in making my choice. So here we go.

The thing I most want everyone to know about the human brain is that it is pretty unique; its awesome by-products—attributes such as the sense of self, creativity, and intuition—and its ability to create mental abstractions (art, mathematics, myths, and science), which are used in our attempts to define and describe material reality, and tools that increase our reach into the world are all beyond anything we have ever encountered in the vast universe that surrounds us. So exceptional is our brain that in deference to what it has been able to accomplish during the past hundred thousand years of our species' history, I like to refer to it as "the True Creator of Everything."[10]

A second thing I would like everyone to know most about the human brain is that neither the human nervous system nor its most exquisite products—things such as intelligence, intuition, creativity, and empathy—can be reduced to a simple mathematical algorithm. This kind of "copyright protection," as I like to describe it, is ensured by a series of evolutionary, neurobiological, mathematical, and computational constraints that cannot be overcome either by software, like that proposed by modern artificial intelligence, or by hardware, as the classic proponents of cybernetics once believed.[11] Essentially, all these constraints assure that our brains work through a multitude of noncomputable functions that are beyond the reach of any universal Turing machine, and it means that no digital computer, no matter how sophisticated and powerful it is, will ever be able to reproduce the type of brain we carry between our ears. One reason is that complex brains like ours combine both analog and digital types of processing in their routine operation.[12] In fact, Ronald Cicurel and I[13] have proposed that it is the recurrent and highly dynamical, nonlinear interaction between such analog and digital neuronal processes that creates the sort of functions that endow animal brains with a degree of complexity that far exceeds the capability of any Turing machine.[14] A full account on how this analog-digital model of brain function operates can be found elsewhere.[15] Essentially, we propose that key brain functions are noncomputable by a typical digital computer, no matter how complex it is.

Having quickly disclosed a fundamental reason why I believe that complex animal brains, including ours, operate in a unique way, let me now justify why I refer to the brain as the true creator of everything. Thanks to the neurobiological properties of our brains, we are capable of creating a coherent description of material reality—all that is out there around us. The only description we can have of "what is out there" is the one continuously sculpted by the complex circuits that connect the close to eighty-six billion neurons that form a typical human brain. Even our notion of the basic parameters that define the scaffolding of this external universe, space and time, are perceived in a very particular way by all of us, thanks to the way in which our brains operate (see essay by Hussain Shuler and Namboodiri in this volume). This view of the brain goes against some of the classic textbook depictions of the human nervous system as a passive "decoder" of information provided by the external environment. Rather, based on thirty years of accumulated neurophysiological evidence, collected in my lab and many others, the brain can only be described as a "creator" that defines everything we experience based on "its own internal point of view."[16] Indeed, I like to say that building expectations of what the future may bring—in the next few hundred milliseconds or years ahead—is a key function of our brain since this allows it to establish an internal model of reality, through which it evaluates a given situation, looking for things that match, or not, its original internal model; we see before we watch and hear before we listen.

Our brains are always a step ahead of what is about to happen. In case a mismatch is identified, our brains also have the exquisite ability to learn from these mistakes and quickly update their internal models of reality accordingly. This learning is mediated by a more general property known as neuronal plasticity; it is the ability of our brain to reorganize its functions, and even its physical microanatomical structure, as a result of a novel experience or changes in the surrounding environment (see essays by Wilbrecht, Lau and Cline, and Barth in this volume). The degree of brain plasticity, even in adult animals, can truly relate the human central nervous system to a kind of orchestra, one in which every note produced is capable of altering the physical configuration of the instruments that contributed to the generation of that note. As such, the brain that listens to the last sound in a symphony is not the same that existed when the symphony began to play.

Brains like ours are also continuously exchanging information across multiple levels of organization, from the molecular to the cellular to the circuit levels, in an operation that involves the continuous and instantaneous updating of a number of parameters so large that it is difficult to even find a word to describe the magnitude of this process. All this complexity allows the human brain to generate all the attributes that define the human condition: the entirety of our culture, history making, and civilization building; our incomparable prowess for tool making and technology development; our special skills to communicate through language or even to create a huge variety of artificial media that allows us to establish enduring social groups over vast spans of time and space; our artistic manifestations and scientific conquests; our ethical and religious beliefs; these are all part of the immense catalog of noncomputable realizations that take place as a result of the type of neurobiology that governs our central nervous systems.

Although I am convinced that we can dismiss the possibility that a digital version of the human brain will ever be built, I would like to raise a much more concrete and troublesome scenario: the possibility that, as a result of overexposure to digital systems, our brains may, through the process of neuronal plasticity, began to mimic the operation and logic of these digital systems, simply because of the considerable rewards offered by emulating this type of machinelike behavior.[17] In their books, Nicholas Carr and Sherry Turkle offer a glimpse of this potential and (at least for me) unwelcome future.[18] They describe a variety of cases in which our overindulgence with digital systems, including social media, may be affecting some of our key brain functions. From radiologists who reduce their ability to diagnose certain images due to an overreliance on automatic image recognition software, to a sense that architectural creativity is being reduced by the use of design software by big architecture firms, to the increase in young adults' anxiety due to an overwhelming sense of loneliness experienced by those who spend a large fraction of their days interacting virtually on social media, the initial signs are everywhere. At the limit, delegating our intellectual and social tasks to digital systems may just curtail or simply eliminate a variety of unique human behaviors, transforming our brains into mere biological digital systems. Although I see this potential scenario as rather tragic and highly undesirable as a legacy to future generations, I am afraid I cannot dismiss the concrete risks that it may actually become part of our future reality as easily as I can express my awe and amazement of the masterpieces brought to life by the true creator of everything.

NOTES

1. "Meat machine" was the way Marvin Minsky, an MIT researcher on artificial intelligence, once referred to the human brain.
2. C. Shannon and W. Weaver, *The Mathematical Theory of Communication* (Urbana: University of Illinois Press, 1949). Shannon's theory of communication was originally limited to quantify the reliability of messages transmitted over noisy communication channels like undersea cables. In this context, Shannon defines information as a measurement—entropy—that represents the minimum number of bits needed to accurately encode a sequence of symbols, each of which has a particular probability (or frequency) of occurrence.
3. D. M. MacKay, *Information, Mechanism, and Meaning* (Cambridge, MA: MIT Press, 1969).
4. N. K. Hayles, *How We Became Posthuman: Virtual Bodies in Cybernetics, Literature, and Informatics* (Chicago: University of Chicago Press, 1999); P. N. Edwards, *The Closed World: Computers and the Politics of Discourse in Cold War America* (Cambridge, MA: MIT Press, 1997).
5. Ibid.
6. J. Weizenbaum, *Computer Power and Human Reason: From Judgment to Calculation* (New York: Freeman, 1976); R. Penrose, *The Emperor's New Mind: Concerning Computers, Minds, and the Laws of Physics* (Oxford: Oxford University Press, 1989); J. Searle, *The Rediscovery of the Mind* (Cambridge, MA: MIT Press, 1989).
7. A. Turing, "On Computable Numbers, with an Application to the Entscheidungsproblem," *Proceedings of the London Mathematical Society Series 2* 42 (1936): 230–265.
8. Ibid.
9. https://plato.stanford.edu/entries/church-turing/

10. M. A. L. Nicolelis, *The True Creator of Everything: How an Organic Computer—the Human Brain—Created the Universe As We Know It, and Why It Is Now under Attack by Some of Its Most Powerful Creations* (New York: Basic Books, forthcoming).

11. Cybernetics was a movement created in the late 1940s by, among others, the MIT mathematician Norbert Wiener. It tried to reduce all the physiological operations that take place in a human body, particularly inside the human brain, to a collection of information processes. In this context, members of the cybernetics movements believed they could analyze a human being inside a control-loop system—like that of a radar-based, anti-aircraft artillery system—as if the human being behaved like an automaton.

12. Analog neuronal signals vary continuously in time. They can be generated by the average of the electrical/magnetic activity of large numbers of neurons, like the electroencephalogram or magnetoencephalogram, or by signals generated by a single neuron, like the action potential and synaptic potentials. A digital neuronal signal can be obtained by recording only the timing in which single neurons fire action potentials.

13. R. Cicurel and M. A. L. Nicolelis, *The Relativistic Brain: How It Works and Why It Cannot Be Simulated by a Turing Machine* (Natal, Brazil: Kios Press, 2015).

14. Turing, "On Computable Numbers."

15. Cicurel and Nicolelis, *The Relativistic Brain.*

16. M. A. L. Nicolelis, *Beyond Boundaries: The New Neuroscience of Connecting Brains with Machines and How It Will Change Our Lives* (New York: Times Books, 2011).

17. M. A. L. Nicolelis, "Are We at Risk of Becoming Biological Digital Machines?" *Nature Human Behavior* 1 (2017); DOI: 10.1038/s41562-016-0008.

18. N. Carr, *The Glass Cage: Automation and Us* (New York: W. W. Norton, 2014); S. Turkle, *Alone Together: Why We Expect More from Technology and Less from Each Other* (New York: Basic Books, 2011).

There Is No Principle That Prevents Us from Eventually Building Machines That Think

Michael D. Mauk

ARTIFICIAL MINDS are all the rage in books and movies. The popularity of characters such as Commander Data of *Star Trek Next Generation* reveals how much we enjoy considering the possibility of machines that think. This idea is made even more captivating by real-world artificial intelligence successes such as the victory of the Deep Blue computer over chess champion Garry Kasparov and the computer Watson's domination of the TV game show *Jeopardy*. Yet while victories in narrowly defined endeavors like these games are truly impressive, a machine with a humanlike mind seems many years away.

What will be required to span the gap is surprisingly simple: mostly hard work. To date, no fundamental law has emerged that precludes the construction of an artificial mind. Instead, neuroscience research has revealed many of the essential principles of how brain cells work, and large-scale "connectome" projects may soon provide the complete wiring diagram of a human brain at a certain moment in time. Yes, there are many details left to discover, and to realize this goal the speed and capacity of computers must grow well beyond today's already impressive levels. There is, however, no conceptual barrier—no absence of a great unifying principle—in our way (but see the Nicolelis essay in this volume for a counterview).[1] Here, my goals are to make this claim more concrete and intuitive while showing how research attempting to recreate the processing of brain systems helps advance our understanding of our brains and ourselves.

Pessimism about understanding or mimicking human brains starts with a sense of the brain's vastness and complexity. Our brains are comprised of around eighty billion neurons interconnected to form an enormous network involving approximately five hundred trillion connections called synapses. As with any computing device, understanding the brain involves characterizing the properties of its main components (neurons), the nature of their connections (synapses), and the pattern of interconnections (wiring diagram). The numbers are indeed staggering, but it is crucial that both neurons and their connections obey rules that are finite and understandable.

The eighty billion neurons of a human brain operate by variations on a simple plan. Each generates electrical impulses that propagate down wirelike axons to synapses, where they trigger chemical signals to the other neurons to which they connect. The essence of a neuron is this: it receives chemical signals from other neurons and then generates its own electrical signals based on rules implemented by its particular physiology. These electrical signals are then converted back into chemical signals at the next synapse in the signaling chain. This means that we can know *what* a neuron does when we can describe its rules for converting its inputs into outputs—in other words, we could mimic its function with a device that could implement the same sets of rules. Depending on how fine-grained we make categories, there are on the order of hundreds (not tens, not thousands) of types of neurons.[2] So determining a reasonably accurate description of the input-output rules for each type of neuron is not terribly daunting. In fact, a good deal of progress has been made in this regard.[3]

What about the synaptic connections between neurons? There do not appear to be any barriers that can prevent our understanding these structures. The principles governing how they work are increasingly well understood. Synapses contain protein-based micromachines that can convert electrical signals generated by a neuron into the release of miniscule amounts of chemical neurotransmitter substances into the narrow gap between two connected neurons. The binding of neurotransmitter molecules to the receiving neuron nudges the electrical signals in that neuron toward more spiking activity (an excitatory connection) or toward less spiking activity (an inhibitory connection). There remain many de-tails to discover about different types of synapses, but this task is manage-able with no huge conceptual or logistical barriers.

When appropriate, our understanding of a synapse type would have to include its ability to persistently change its properties when certain patterns of activity occur. These changes, collectively known as synaptic plasticity, can make a neuron have a stronger or weaker influence on the neurons to which it connects. They mediate learning and memory—our memories are stored by the particular patterns of the strength of the tril-lions of synapses in our brains. It is important that we need not know every last molecular detail of how plasticity works. To build a proper ar-tificial synapse we simply need to understand the *rules* that govern its plasticity.

Even with eighty billion neurons then, the number of neuronal *types* and the number of plasticity *rules* is finite and understandable. It is an entirely doable task to build artificial versions of each neuron type, whether in a physical device such as a chip or as a software subroutine. In either case, these artificial neurons can receive certain patterns of input across their synthetic synapses and can return the appropriate patterns of output. As long as each artificial neuron produces the correct output for any pattern of inputs its real version might receive, we have the build-ing blocks needed to build an artificial brain.

With the vast complexity of each brain's wiring diagram it's easy to imagine there is just too much complexity to understand, even with a manageable number of neuron types. However, there are reasons to be-lieve that the complexity can be tamed. The connections between neurons are not random. Rather, they follow rules that are readily identifiable. For example, in a brain region called the cerebellar cortex, a type of neuron called a Golgi cell receives synapses from many axons called mossy fi-bers and also from the axons of granule cells. The Golgi cells can also, as is turns out, extend axons and form synapses to inhibit their neighbor Golgi cells (it's reciprocal too: they are inhibited by their neighbors). With this information we could wire up the Golgi cells of an artificial brain.

There are important differences between building a brain and build-ing a particular person's brain. We need to know all of the specific con-nections only if we are trying to build a replica of a particular person.

Your specific connections are different from mine—that's why we behave differently, why we have different reactions and predilections, and especially why we have different memories. But the connections of your brain and of my brain were produced, starting in utero and continuing throughout life, following the same sets of rules, together with individual experience. If we wanted to make a replica of your brain or of mine, we would need to know about, and be able to reproduce a map of, all of the hundreds of trillions of specific connections and their associated properties. If, however, we simply want to build a replica human brain that will have its own reactions and predilections, then we need only follow the basic rules to produce connections.

The same arguments hold for synaptic plasticity and memories. If we want to replicate a particular human with his or her particular memories, we have to correctly get not only all of the specific connections, but also the strength of each connection. If we want to build another human that, over time, will develop his or her own memories and personality, then we only need to build in the proper rules for plasticity at each kind of synapse. This is a vastly more tractable task than the specific, synapse-by-synapse information that would be needed to build an artificial version of a particular person (at a particular time in that person's life).

Even after we appreciate how the vast complexity of the brain can be broken down into manageable bites, it can still be difficult to wrap our heads around how all of this might relate to behavior, cognition, emotion, and memories. This is where engaging in thought experiments can be helpful. They are thought experiments because we can't actually do them because they are either not feasible or affordable or whatever. But just the process of imaging them and the outcomes they might have can be illuminating. Einstein famously used thought experiments about what it would be like to travel on a beam of light to develop his ideas of special relativity. Fortunately, we don't have to be Einstein to make use of thought experiments about how brains think.

It's best to start simple, so let's consider a basic reflex, such as an eye blink, which results when something touches the skin near your eye. The touch causes activity in sensory neurons with endings in the skin that send their axons to the brain stem, where they activate other neurons through excitatory synapses. Those neurons in turn activate yet another type of neuron, called motor neurons, which project to the muscles of the eyelid, causing the reflexive blink. Imagine that we had the technical ability to build artificial devices that are small enough to substitute for each of these neurons in the reflex. They would receive inputs and produce outputs according to the very same rules as the three types of neurons in this reflex. Once implemented, the reflex would behave in precisely the same way as it did with real neurons: touching the skin would produce reflexive blinks that would be indistinguishable from normal blinks.

When we see something, the underlying neural processes are essentially the same as the eye blink reflex, only vastly more complex. The photoreceptor cells in the retina absorb patterns of light and, through a network of neurons in the retina, convert them to electrical signals that are eventually sent via the optic nerve to the brain. There, several layers of neurons are activated or inhibited, according to the pattern of inputs they receive and to the rules each neuron uses to convert inputs to outputs. These outputs, conveyed by axons and synapses, then become the inputs to other parts of the brain where neuron activity controls things such as perception of and reaction to the visual scene. If we applied the same artificial neuron technology to this vastly more complex network, the results would be the same. For a given visual scene, our perception would be the same after replacement of biological neurons with their synthetic counterparts. The patterns of activity would be the same, and the outputs would be the same, even if the underlying processes that produced the activity in the artificial neurons were entirely unlike those of natural neurons.

Now imagine that our technology weren't quite as advanced. We can build artificial neurons, but we can't make them small enough to fit in their place in the brain. What we can do instead is build devices that can sense the release of neurotransmitters (from real neurons or other artificial ones), and they can release transmitters as output to other neurons, real or artificial. However, imagine that the processing of the rules for when outputs are produced given a particular set of inputs can't be squeezed into this physical device. So this processing has to be done in a computer, with signals wirelessly traveling back and forth between this computer and the implanted devices via some super-advanced Bluetooth interface. This thought experiment reveals that even with the actual processing disembodied in this way, the flow of information through the network could still be the same. Every fake neuron would be activated in precisely the same patterns as those of the real neurons.[4]

In this way, if we could substitute for each neuron in a brain an artificial device that implements the same rules at the same speed, there is no reason to expect that the behavior of the person with such a brain, or the conscious awareness of that person, would be any different. Moreover, any physical computing device that can be built can be implemented in a computer program. This means that, eventually, we will be able to perform the greatest experiment ever—translating what we know about the brain into the production of a computer program that implements a mind. It is crucial that it would take more than a brain, though. Asif Ghazanfar's essay in this volume is an important reminder that brains must be embodied so that they get the rich and vast arrays of inputs that our bodies send to our brains.

Back down from our rarified thought experiment world, neuroscientists are already building computer simulations of different parts of the brain. They build subroutines that implement the input/output rules of real neurons. These subroutines are set aside in portions of the computer's memory. To simulate a brain system it is necessary to set aside many portions of memory, each devoted to a different copy of that type of neuron. The connectivity between these neurons is implemented by storing in these portions of memory the identity of the other neurons to which each neuron connects. That way, each time an artificial neuron produces an output, the portions of memory holding the downstream neurons can be informed so that the processing of those neurons' activities can be updated. Such simulations can be enormously useful to neuroscientists. They help identify what we don't know when they fail and help us understand how things work and to think of the most essential new experiments when they work properly. In this way, experiments can inform simulations and make them better, even while simulations can inform the design and interpretation of experiments.

We are probably many years away from building artificial minds. And along the way we will have to grapple with ethical issues about how we treat these creations of our genius. But unlike quantum gravity or space flight to other galaxies, there are no theoretical barriers in our way.

NOTES

1. For another counterview, see R. Penrose, *The Emperor's New Mind: Concerning Computers, Minds, and the Laws of Physics* (Oxford: Oxford University Press, 1989).
2. Neuron types vary quite a bit in their processing and activity. Some are mostly silent in the absence of inputs from other neurons, while others fire spontaneously. Some neurons when excited fire only at the onset of an input, and others fire continuously.
3. G. M. Shephard, *The Synaptic Organization of the Brain*, 5th ed. (Oxford: Oxford University Press, 2004).
4. In some ways, the differences between the real brain and an artificial mind wouldn't be all that great. The electrical signals in the brain's neurons are produced by ions flowing through channels, which are a form of gate, in the neurons. This current flow produces voltage changes that implement the brain's signal processing. In the processor of a computer—the one running our artificial brain program—the signals are produced by electrons flowing through transistor gates. In both instances, electrical signals undergo computational transformation following a specified set of rules.

Epilogue

I BEGAN THIS BOOK by asking thirty-five neuroscientists (or pairs of neuroscientists) a simple question: "What idea about brain function would you most like to explain to the world?" Their answers reflect their delightful individual proclivities and quirks. Of course, if I had asked a different group of thirty-five, a different book would have come about. However, even though this collection is, by design, incomplete and serendipitous, I believe that some overarching themes about the function of the nervous system have emerged:

Most complex behavioral traits, like shyness or sexual orientation or novelty-seeking, result from a mixture of genetic and environmental factors and the interplay between them. Environmental factors include both biological factors, like the hormones to which one is exposed in utero and early life, and our experiences in the world, including social interactions. At a molecular level, one crucial way that experience changes the nervous system is by turning gene expression (the means by which a gene instructs the production of a protein) on or off in neurons.

The brain is molded by experience in different ways during development, throughout life and in response to injury. Wiring up the brain, with its many connections, is a complex task that cannot be entirely directed by genetics. The solution appears to be twofold. First, some very simple rules can be used to direct very particular neural wiring. Second, the fine details of wiring and certain electrical properties of neurons and synapses are guided by experience. This *neural plasticity* is crucial for wiring up the brain during fetal and early postnatal life. Neural plasticity persists throughout life and is the basis for memory and some aspects of human individuality. It is crucial to adapting our brains and nervous systems to our changing bodies across our lifespan and in response to bodily or neural injury. Neural plasticity also allows us to change our brains rapidly in response to changes, like tool use, and slowly but profoundly in response to repeated training, as in athletics or musical performance.

Our experience of the world (and of our own bodies) is actively constructed by the brain. The nervous system is not built to give us the most accurate representation of the external world. Rather, it gives us a view of the world that, in the past, has been the most useful for surviving and getting our genes into the next generation. There are no pure sensations. Our brains are built to cherry-pick and emphasize certain aspects of the sensory world and then to blend those sensations with emotions and expectations. We're hardwired to pay attention to changes in our environment and ignore sensations that persist. Our perception of time is warped by our anticipation of reward. Ultimately, all perception is in the service of useful decision and action, not objective accuracy.

Much of what the brain does occurs below the level of conscious awareness. For example, taste receptors in the mouth not only contribute to our sensation of flavor, but also prepare our digestive system for what is about to arrive. With practice, even complex actions, like driving to work, become habits and fade from our active thoughts. A large portion of the brain is concerned with predicting what will happen in the very next few moments, and these subconscious calculations do not require us to attend to them.

Our hominin ancestors have lived in social groups for a very long time, and this has led us to be exquisitely sensitive to subtle interpersonal cues. These cues include facial expression, tone of voice, direction of gaze, and other forms of body language. They contribute to our crucial ability to infer what another person knows or might be thinking, as well as his or her emotional state.

We feel like autonomous and entirely rational beings, but we are all subject to strong subconscious drives and motivations, most of which are related to survival and reproduction. These drives impact not only our romantic and sexual desires, but also our response to such high-level culturally constructed phenomena as art and advertising.

When I was starting college in the late 1970s, back in the stone knives and bearskins era of neuroscience, I was given explanations about brain function that have since been shown to be wrong. Some of these explanations were absolutely 180 degrees dead wrong. For example, I was taught that each neuron in the nervous system could release only one type of neurotransmitter, like glutamate, serotonin, dopamine, or GABA. This idea has now been shown to be false; many neurons can release two or more transmitters and can even do so at the same synapse.[1] Knowing that neurotransmitters can be co-released has important implications for the function and evolution of neural circuits and is also crucial to those who seek to create effective neuropsychiatric drugs.

More often, what I was taught in those early days was not dead wrong but rather half right. In 1978, I was told by well-meaning professors that a neurotransmitter was released from nerve terminals when an electrical spike invaded, thereby opening voltage-sensitive calcium channels, allowing Ca ions to rush in. These Ca ions, which have a positive charge, were said to bind to negative charges on the inner surface of the presynaptic membrane and outer surface of the transmitter-laden vesicle, thereby dissipating the electrical repulsion between two negative charges and allowing the vesicle to fuse with the presynaptic membrane.[2] This fusion event then released neurotransmitters into the synaptic cleft. This explanation was half right in that spike-triggered Ca ion influx really is a trigger for vesicle fusion and neurotransmitter release. But it was also wrong because Ca ion binding to the vesicular and presynaptic membranes to screen opposing negative charges is not the trigger for vesicle fusion. Now we know that there are specialized Ca-sensing molecules called synaptotagmins embedded in the membrane of vesicles. Synaptotagmins bind the Ca ions that rush in, and these Ca-bound synaptotagmin molecules allow for vesicle fusion and neurotransmitter release by forming a complex with another group of proteins in the vesicle and the presynaptic membrane. In this case, the general idea about Ca influx as the trigger for neurotransmitter release was right, but the molecular details were all wrong. Furthermore, we now know that there are exceptions to this explanation. Certain neurotransmitters, like nitric oxide, are diffusible gases that readily penetrate cell membranes. These so-called gasotransmitters are formed as needed by a Ca-triggered chemical process and are never stored in vesicles. In this way they completely bypass the requirement for Ca-bound synaptotagmin molecules and their binding partners (see the Snyder essay in this volume).

There is no question that in time, some of what you have read here will turn out to be half right and some explanations may even be dead wrong. That's not because of carelessness; it's because of the essential nature of scientific inquiry. As William Kristan and Kathleen French so eloquently explore in their essay, science is an ongoing process in which every hypothesis is subject to testing, reformulation, and refinement. Like evolution, its work is never done. No individual or species is optimally adapted to its environment, and no scientific idea is immune to challenge, testing, and refinement or even complete rejection. Belief in the scientific endeavor is belief in the process, not in the absolute truth of our present understanding.

A great example of the ongoing scientific process may be found in the two forward-looking essays that close this volume. Michael Mauk has written that there is no principle to be found in neuroscience that will prevent us from ultimately building machines that think. He holds that all we need to do to make thinking machines is a lot of hard work to understand behavior as well as the wiring and signals in the brain, coupled with faster computers and larger storage devices. On the other hand, Miguel Nicolelis proposes that the interaction of digital and analog signals, as well as some other factors in the design of the human brain, ensures that our brains work with uncomputable functions that cannot be solved by any future computing machine, no matter how powerful or capacious.

Who's right about this crucial point? We don't know. If the past is any guide, the answer will be complicated and will reveal that the question was not ideally stated at the outset. That's not a critique of the careful and thoughtful Drs. Nicolelis and Mauk. It's just the way things tend to progress in science. At any moment in time we can only imagine enough to formulate a certain range of questions. With time and effort, that range of questions shifts—sometimes slowly, sometimes quickly.

NOTES
1. In some cases two different neurotransmitters are even loaded into the same vesicle.
2. Both vesicles and cell membranes are composed of a phospholipid bilayer in which the negatively charged head groups of the phospholipid face outward. It is these negatively charged head groups that attract the positively charged Ca ions that rush into the presynaptic terminal.

CONTRIBUTORS

SCOTT T. ALBERT, Johns Hopkins University School of Medicine

ALISON L. BARTH, Carnegie Mellon University

ALLAN BASBAUM, University of California, San Francisco

AMY BASTIAN, Kennedy Krieger Institute and Johns Hopkins University School of Medicine

PAUL A. S. BRESLIN, Rutgers University and Monell Chemical Senses Center

LUCY L. BROWN, Albert Einstein College of Medicine

ANJAN CHATTERJEE, University of Pennsylvania

HOLLIS CLINE, Scripps Research Institute

CHARLES E. CONNOR, Johns Hopkins University

ANIRUDDHA DAS, Columbia University

GÜL DÖLEN, Johns Hopkins University School of Medicine

DAVID FOSTER, University of California, Berkeley

KATHLEEN A. FRENCH, University of California, San Diego

ASIF A. GHAZANFAR, Princeton University

DAVID D. GINTY, Harvard Medical School and Howard Hughes Medical Institute

ADRIAN M. HAITH, Johns Hopkins University School of Medicine

JULIE KAUER, Brown University

DARCY B. KELLEY, Columbia University

ALEX L. KOLODKIN, Johns Hopkins University School of Medicine

JOHN W. KRAKAUER, Johns Hopkins University School of Medicine

WILLIAM B. KRISTAN, JR., University of California, San Diego

MELISSA LAU, University of California, San Diego

DAVID J. LINDEN, Johns Hopkins University School of Medicine

LIQUN LUO, Stanford University

PEGGY MASON, University of Chicago

MICHAEL D. MAUK, University of Texas at Austin

CYNTHIA F. MOSS, Johns Hopkins University

VIJAY M. K. NAMBOODIRI, Johns Hopkins University School of Medicine

JEREMY NATHANS, Johns Hopkins University School of Medicine

MIGUEL A. L. NICOLELIS, Duke University

YAEL NIV, Princeton University

MICHAEL PLATT, University of Pennsylvania

INDIRA M. RAMAN, Northwestern University

TERRENCE SEJNOWSKI, Salk Institute, Howard Hughes Medical Institute, and University of California, San Diego

REZA SHADMEHR, Johns Hopkins University School of Medicine

MARSHALL G. HUSSAIN SHULER, Johns Hopkins University School of Medicine

SOLOMON H. SNYDER, Johns Hopkins University School of Medicine

SCOTT M. STERNSON, Janelia Research Campus, Howard Hughes Medical Institute

SAM WANG, Princeton University

LINDA WILBRECHT, University of California, Berkeley

ACKNOWLEDGMENTS

First and foremost, my profound thanks to the contributors for their splendid essays, enthusiasm, and good will. It takes a leap of faith to sign onto an odd project like this one, and I'm extremely grateful to all who jumped into the pool. I hope, as you're toweling off, that you're as happy with the outcome as I am. And a special shout out to John Krakauer and Sascha du Lac, who helped to formulate the list of contributors.

The wonderful members of the crew in the Linden Lab have been sounding boards for many of the ideas herein and have been gracious in putting up with my distraction. Special thanks to Devorah Vanness, Michelle Harran, and Jessie Benedict for going above and beyond the call of duty—thanks for your watchful eyes and steady hands.

And, of course, a big tip of the hat to the publishing pros—Jean Thomson Black, Michael Deneen, Ann-Marie Imbornoni, and Bojana Ristich at Yale University Press and Andrew Wylie, Jackie Ko, and Luke Ingram at the Wylie Agency for their warm advocacy and guidance.

INDEX

Note: Page numbers in italics refer to illustrations.

memory formation, 66–71

mentalizing. *See* Theory of Mind

Menzel, Randolph, 259

Merkel cells, 123–124, 126

metabolic priming, 111–112

metabotropic receptors, 5

micronutrients, 112, 117n7

midbrain reward regions, 210, 211

mind attribution. *See* Theory of Mind

mind reading. *See* Theory of Mind

Minnesota Study of Twins Reared Apart (MISTRA), 21–23

MIT, 106–107

monkeys. *See* primates, nonhuman

Montague, Read, 259

Moser, Edvard and May-Britt (et al.), 148, 157

Moss, Cynthia F., 153–160

motherese, 187

mother's caring behaviors, 201–203, 205n2, 205n4

motivation (dopamine-releasing neurons), 257–262

motivation (the will), 245–250

motivational circuit, 68–70

motivational need states, 248

motor cortex, 54–55, 85, 162, 170–172

motor learning: error-based, 163–166; and proficiency, 41–44; and stroke patients, 168–169

Mountcastle, Vernon, 88

mouth, taste receptors in, 112–113

musicians and cortical changes, 54–55

Nadel, Lynn, 147

Namboodiri, Vijay M. K., 135–144

Nathans, Jeremy, 19–25

National Institutes of Health (NIH), 88, 89

nausea, gagging, and vomiting, 113, 115–116

Nav1.7 subtype of sodium ion channel, 131–132

necessity (test for causality), 11

negative reinforcement, 246–250

neocortex, 6, 61, 211, 257

nervous system: importance of body shaping and guiding, 254–255; as passive decoder of information, 266. *See also* neural connections

neural connections: and brain simulation, 272–273, 275; childhood plasticity, 41–44; difficulties of unique label strategy of, 26; and experience, 277–278; and fruit fly eyes, 28–31, 29; genetics and environment and, 7–8; Ramón y Cajal and, 27; and the teenage brain, 46–49

neural plasticity. *See* plasticity

neural superposition, 29, 31

neurobullshit, 10

neuronal firing patterns, 144n14

neurons: basic structures, 2–3, 3; behavioral functions of, 15n6; and brain simulation, 271; a century of discoveries about, 145–150; first appearance of, 2; flow of electrical information, 3, 3–5; hyperactive, 80n5; interconnections/wiring (*see* neural connections); replacement of damaged cells, 32, 33n10

neurotransmitter receptor proteins, 77–79, 80n6

neurotransmitter receptors, 4–5, 78–79, 80n6

neurotransmitters: and brain simulation, 271–272; characteristics of, 3, 88, 91–92, 279–280; discoveries of, 88–91

Nicolelis, Miguel A. L., 263–269, 280

nicotine addiction, 68

NIH (National Institutes of Health), 88, 89

nitric oxide, 91

Niv, Yael, 227–230

Nobel Prizes, 36, 89, 112, 146, 148, 155, 157

nociceptors, 129–131

noradrenaline (norepinephrine), 80n4, 89

Novosibirsk study, 20